A TALE OF TWO CAVES

HARPER'S CASE STUDIES IN ARCHAEOLOGY

William A. Longacre,
General Editor

A TALE OF TWO CAVES

FRANÇOIS BORDES

Université de Bordeaux

HARPER & ROW, PUBLISHERS

New York • Evanston • San Francisco • London

A TALE OF TWO CAVES

For information address
Harper & Row, Publishers, Inc.,
49 East 33rd Street, New York, N. Y. 10016.

Standard Book Number:
06-040855-3 (cloth)
06-040854-5 (paper)

LIBRARY OF CONGRESS CATALOG CARD NUMBER:
75-174539

CONTENTS

EDITOR'S FOREWORD

As student interest in anthropology and archaeology has increased in recent years, I have seen a need for a series of supplementary texts such as "Harper's Case Studies in Archaeology." The rapid growth of enrollments in introductory courses and the offering of such courses in ever larger numbers of colleges and universities across the country have placed a considerable strain on the relatively meager resources available, particularly those used in the teaching of archaeology. My concern over the lack of teaching resources in archaeology is shared by many in the profession. I am grateful not only for the positive response to the idea of a series of short monographs reporting problem-oriented archaeological research but also for the commitment of scholars to produce exciting accounts of their current research.

The books in this series are designed to show the methods by which archaeologists solve research problems of broad anthropological significance. The range of problems, time periods, and areas of the world studied has been selected to provide maximum utility and flexibility for the teachers of introductory anthropology and archaeology courses. The series will provide an integrated set of monographs that introduce the student to basic aspects of modern archaeological research. There is probably no more effective means to convey the excitement and relevance of modern, problem-oriented archaeological investigation than to have the investigators who conceive and carry out such research prepare the monographs themselves. By providing an integrated view of the most recent directions in archaeological research, the monographs will challenge the instructor to put recent investigations into historical perspective and encourage the student to develop an appreciation of the importance of the development of anthropological archaeology in light of the present thrust of our discipline. The student will also share in the current emphasis upon relevant research designed to reveal the nature of cultural evolution. The emphasis of the series is explicitly upon process—processes of change and stability in the development of culture. As such, the series is designed to contribute to the teaching of modern anthropology and to emphasize the important role that archaeological research plays in achieving the larger goals of anthropology.

THE AUTHOR AND HIS BOOK

François Bordes is Professor and Director of the Institute of Prehistory, University of Bordeaux, France, and Director of Prehistoric Antiquities in southwestern France. He received the Doctor of Science degree from the University of Paris in 1951 and today is recognized throughout the world as a leading authority on the Paleolithic cultures of Europe. He has been a Visiting Professor at numerous institutions in the United States and Canada, including the Universities of Arizona, California, and Chicago.

His special interests and research have centered on the prehistory and Pleistocene geology of southwestern France. His immense contributions to our understanding of cultural developments during the various Paleolithic periods in France are the result of his vigorous and highly innovative research.

Among his innovations are refinements in techniques of excavation, more precise determination of the location of artifacts, a focus upon "living floors" rather than upon geological strata in excavation, and great contributions to the analysis of excavated material. In this last category, he has applied his vast knowledge and experience in the technology of stone toolmaking to refine the typology of flint artifacts. This, along with his application of various statistical techniques to the distribution of types of tools in Paleolithic sites in France, has produced a far more detailed understanding of cultural developments during the Paleolithic than was possible before. His new understanding of the Paleolithic and his innovative research are reflected in this book.

The book focuses upon the excavations of two Paleolithic cave sites in southwestern France: Pech de l'Azé and Combe-Grenal. Professor Bordes outlines the problems that focused attention on these two sites and discusses the historical development of the excavations. In the process, he explains the development and results of the new techniques of excavation and analysis that he has introduced to the study of Paleolithic cultures.

He also introduces us to a most stimulating controversey regarding the interpretation of cultural variants or industries of the middle Paleolithic period or Mousterian culture. Professor Bordes feels that the different Mousterian industries reflect different traditions of tool-making that coexisted in the same area. Two American archaeologists, Lewis and Sally Binford, have presented a different interpretation. The Binfords argue that the differences in Mousterian industries represent different kinds of activities that were carried out by groups participating in a single cultural tradition. Professor Bordes discusses these alternative hypotheses in this book. The interested reader is encouraged to consult Bordes' and D. de Sonneville-Bordes' article, "The Significance of the Variability in Paleolithic Assemblages," which appeared in the June 1970 issue of *World Archaeology*, and two articles by the Binfords, "A Preliminary Analysis of Functional Variability in the Mousterian of Levallois Facies," which appeared in *American Anthropologist* in 1966, and "Stone Tools and Human Behavior," which was published in *Scientific American* in 1969. These alternative hypotheses will stimulate a great deal of future research in Paleolithic archaeology.

William A. Longacre

Introduction

This is a tale of two caves, but of three sites, since one of them is a double site. I shall try to describe how they were discovered and excavated at different periods, under what conditions, and the results and interpretations that followed. Not all the questions will receive an answer. Neither of the caves is yet completely published, nor even completely studied; work on both is still going on. Some of the answers are probably beyond our present reach, since we would need other caves excavated by the same, or better, methods for comparison. This is one of the reasons why, at each site, a large part was left unexcavated. In so doing, we may have lost quite a lot of information, but hopefully we left much more for future archaeologists, who will have the chance to verify our results with techniques we cannot yet dream of.

These two caves have been linked with the development of prehistory in several ways. First, they are among the oldest to have been, not "excavated" in the modern sense of the word, but explored and tested. Second, they rank among the most complex and interesting Mousterian and Acheulean sites in all western Europe. Then, too, the recent excavations, those since 1948, have been largely international: not only did French scientists and students participate but also many foreigners. Archaeologists wanting a firsthand knowledge of prehistory in the "classical region" and many students came to take part—we had with us a large number of Americans and quite a few English, Belgians, Germans, Poles, Swedes, Danes, Israelis, Nigerians, Japanese, Turks, Mexicans, Peruvians, Ecuadorians, Canadians, and Koreans. Czechoslovakian and Russian archaeologists, as well as Yugoslavians and Hungarians, have seen the sites under excavation. And, in addition to all these professionals and students, some amateurs came to help and to learn.

Most of the professionals were archaeologists, or, as we say in France, prehistorians. Others were geologists, paleontologists, pedologists, physical anthropologists, palynologists, sedimentologists, physicists, or chemists. These excavations can be said to have been truly international and interdisciplinarian. And this abundance of people, competences, and efforts has given us a rather clear picture of the chronology, archaeology, environment, ecology, climatic variations, and so on, of these sites, from the penultimate glaciation to the end of the first half of the last glaciation.

It is important to define some of the words that will be used throughout this book. The penultimate glaciation will be termed "Riss," without reference to any special moraine front. And "Würm" will be used for the last glaciation.

As for the subdivisions of these two great cold periods, we shall use the terminology habitually accepted in France: the Riss is subdivided into three stades (even if there are some good indications of a fourth subdivision now) and the Würm into four. If some of you are used to the central European subdivisions of the Würm, let us say that central European Würm I corresponds to our Würm I and II.

We know very well that this is a rough-and-ready classification and that the events were much more complicated, as we shall see later; but it should usefully serve as a first approximation.

Where and When

The southwest of France is a great sedimentary basin lying between the ancient mountains of the Central Massif and the younger Pyrenean range. The shape is roughly triangular, the apex of the triangle leading to the Mediterranean zone via the Naurouze gap, and the base of the triangle being the Atlantic coast. Toward the north, the Aquitaine basin, as it is called, joins the Paris basin via a wider gap, known as the "seuil du Poitou" (Poitou threshold), between the Central Massif and the old Armorican Mountains, which are no more than hills today. To the south, the Pyrenees rise to over 3,000 meters (10,000 feet) (Fig. 1).

Near the base of the triangle, most of the country is covered with Pleistocene gravels and eolian sands. Then come Cenozoic formations, then Upper Cretaceous limestone and sandstone, Jurassic limestone, and finally—on the northern border up against the Central Massif—some Permian sandstone. Thousands of caves or rock shelters exist either in the Mesozoic limestone or in the Permian sandstone. Flint is plentiful almost everywhere, either in the limestone, or, in a secondary position, in the river gravels or Pleistocene formations. Two great rivers irrigate the region: the Garonne, flowing first north from the Pyrenees, and then northwest; and the Dordogne, flowing northeast to southwest. They join a little beyond Bordeaux to form a huge estuary, the Gironde. Both rivers have important tributaries, among them the Vézère, which flows into the Dordogne a little below the village of Les Eyzies.

Today, the climate in the Dordogne district is rather mild, even if there are winter days when the temperature may fall below -17° Centigrade (about 0° F). The mildness is due to oceanic influences, and the area is also marked by rains, mainly in the spring and autumn, but there is comparatively little snow. Of course, during the glacial periods the climate was colder and more continental, but one should not imagine that the Dordogne was ever like Alaska or Labrador. Conditions were not the same

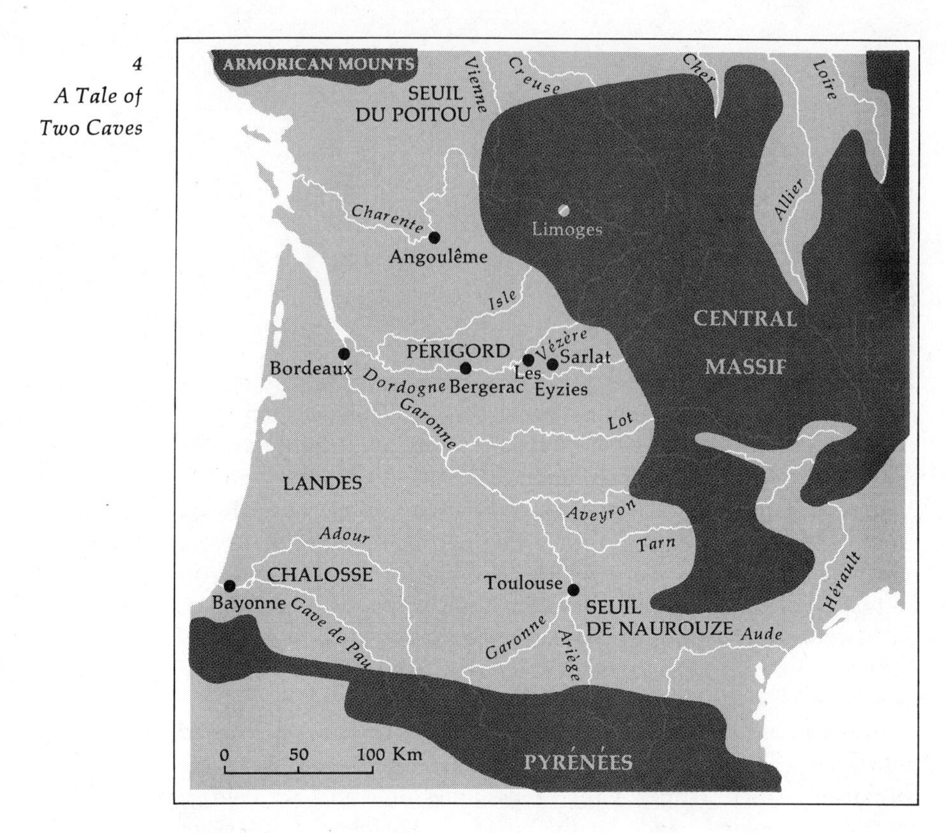

Fig. 1

Map of southwest France.

at all, just as today at about the same latitude there is no comparison between the winters in, say, Bordeaux and Montreal.

The vegetation did vary considerably during the different phases of the Pleistocene, but there was never a true tundra, let alone the Barren Grounds. Plenty of food was always available for the big herbivores.

All these factors explain why the southwest of France was a region of relatively dense prehistoric population. Palaeolithic sites, in caves, shelters, or the open air, run into the thousands, and each year more are found. Here, prehistoric men found flint for their tools, caves and shelters for habitation, a less severe climate than in many other places, rivers to fish, and valleys and plateaux in which to hunt a great variety of animals corresponding to different biotopes. The valleys offered easy means of communication. True, there were small ice caps on the Central Massif, and relatively small ones on the Pyrenees. But seldom was such a set of favorable conditions found by prehistoric man.

Among these various sites, many are well known: La Madeleine and Le Moustier, in the Vézère Valley, and La Micoque, near Les Eyzies, have given their names respectively to the Magdalenian, the Mousterian, and the Micoquian. The Perigordian derives its name from Périgord, an old province of France more or less corresponding to the Dordogne district of today. Aurignac and Mas d'Azil, which gave their names to the Aurignacian and the Azilian, are on the other side of the basin, near the Pyrenees. But close to the village of Les Eyzies are such famous sites as Cro-Magnon, Laugerie-Haute, Laugerie-Basse, and the painted and engraved caves of La Mouthe, les Combarelles, and Font de Gaume, to mention only the best known.

It is only by chance that the two caves studied in this book are less well-known. The first one is called Pech de l'Azé, which, in the local dialect, means "Donkey Hill." The second is called Combe-Grenal, *Combe* meaning a combe or small closed valley, and Grenal (or Grenant) being probably the name of a former owner of the land.

East of Sarlat, a small valley, often dry today, runs southeast to join the Enea, itself a small tributary of the Dordogne River. At least three important sites are situated in this small valley, on the left side: Caminade shelter (Mousterian and Aurignacian), Pech de l'Azé (Acheulean and Mousterian), and Pech de la Boissière (Solutrean and Lower Magdalenian) (see Fig. 2).

Situated about 44°50′ N and 1°15′ E of the Greenwich meridian, the Pech de l'Azé cave is curious in that it has two opposite openings, one facing east, the other west. It traverses a small rocky spur (Fig. 3) and is completely dry today. Indeed, it was already dry during the Acheulean, even though the floor of the valley was certainly higher than it is today.

About 6 kilometers (a little less than 4 miles) south of Pech de l'Azé, on the other side of the Dordogne River, is a small, dry valley situated just east of the medieval town of Domme. In this small valley the cave of Combe-Grenal opens to the south-southwest. Today, what is left is little more than a shelter, but in primitive times it was a small cave. It is situated about 44°45′ N and 1°13′ E of the Greenwich meridian, very close to the Dordogne Valley, and close, too, to the inexhaustible flint supply of the "Plaine de Born," on the plateau. The same flint was exploited in medieval times to make millstones, which were exported by boat on the Dordogne.

All the deposits we are about to study in both caves belong

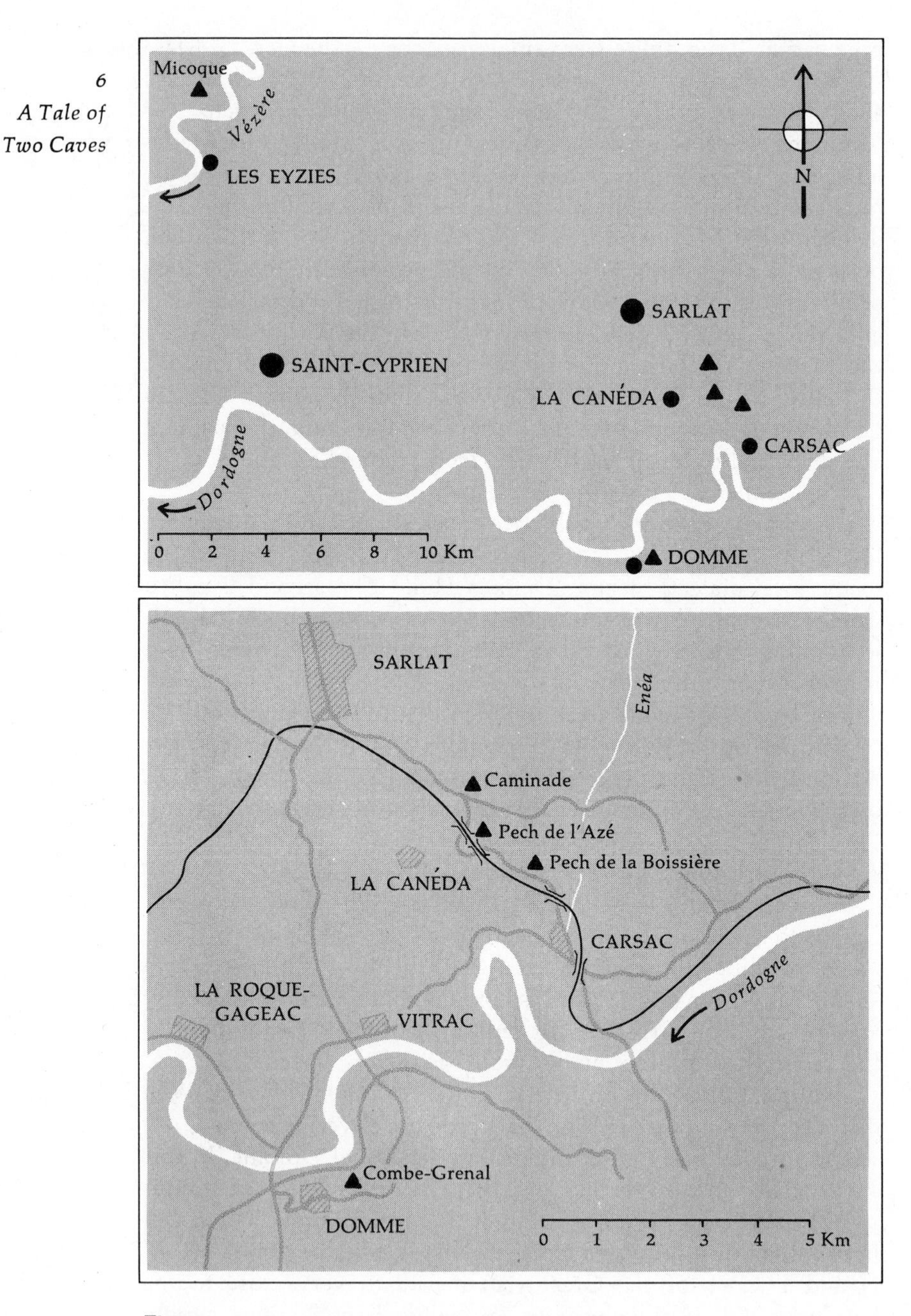

Fig. 2

Top: Map of the country around Sarlat.

Bottom: Enlarged map showing the position of the main prehistoric sites.

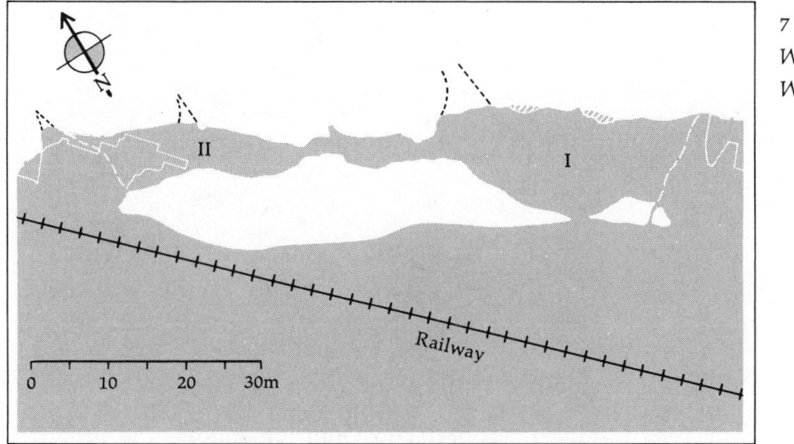

Fig. 3

Plan of Pech de l'Azé Cave. The dotted lines represent the present extension of the roof of the cave at the entrances, or the ramifications of the cave. The fine lines represent the extension of the excavations at both ends.

to the Upper Pleistocene, today thought to begin with the onset of the Riss glaciation. But not far from Combe-Grenal, in the valley of another tributary of the Dordogne (the Céou) lie many caves, some of which have yielded Mindelian deposits, but no implements. Near les Eyzies, in la Micoque, the lowest layer is probably Mindelian. So far, no older human traces have been found in the Dordogne region, but it is very probable that some will be discovered in the future.

It is difficult to give an estimation in years for the pre-Würm deposits. They are beyond the range of radiocarbon dating, and even if suitable material had been found, they are too young for accurate potassium argon dating. Estimations for the beginning of the Würm period vary between 70,000 and 90,000 B.C., the second date seeming the more probable. The Mousterian culture appears to end at about 35,000 B.C.

1. The Pech de l'Azé Cave

THE EARLIER EXCAVATIONS

The road leading from Sarlat to Gourdon follows the little dry valley of an ancient tributary of the Enea. About 3 miles from Sarlat, limestone cliffs with shelters begin to appear on the left. Just where the road forks—one branch going to Souillac, the other to Gourdon—is the Caminade shelter, with several layers of Mousterian and Aurignacian. Then the road curves to the right, goes under a railway bridge, curves back again to the left and under a second bridge. Between the two bridges, on the left, stands Pech de l'Azé hill.

A small path, worn by the feet of hunters and archaeologists, leads up the slope and crosses the railroad tracks. Then there is another slope, which used to be rather steep but was smoothed out in 1969 for the visit of the International Congress on the Quaternary. This leads to the southeast opening of the cave, which is about 50 meters above the dry valley (Fig. 4).

Pech de l'Azé, like so many other caves in this region, was used and partially leveled in medieval times as a sheep pen. So, very probably, some of the deposits were destroyed before any excavation took place. In 1816, a professor of literature who was interested in archaeology, François Jouannet, visited Pech de l'Azé and left a description of it: "The cave is 200 feet long and all stuffed with bones and the teeth of animals. . . ." And in 1818, he added:

> Bones are all over the cave, but mainly at the entrance, where they have been piled up to form a rather thick accumulation. When examined carefully, one can see that they have been partially burned and deliberately broken. With them are mixed, almost in the same proportion, black flints, also broken into small fragments, a circumstance which refutes the idea of natural deposition. . . .

Fig. 4

Top: The Pech de l'Azé hill in winter. The arrow points to the entrance of Pech de l'Azé II. The entrance to Pech de l'Azé I is below the last three pines on the right.

Bottom: The entrance to Pech de l'Azé I. On the front right, wall protecting the section between excavations. Behind, wall built by Professor Vaufrey to contain his back dirt.

Speculating as to when this curious deposit was formed, Jouannet says: "We can see no answer, and if we mention this cave in a chapter on Celtic antiquities, it is only because everything we have seen there bears the mark of the highest antiquity." (18)

A little later, in 1828, another visitor, the abbé Audierne, gave more details:

> The entrance faces the rising sun and is bow-shaped. It is 12 feet high at the center and 35 feet wide. One can see, on both sides, remains of walls where worked flints and bones are found. . . . The cave is horizontal and has another opening to the west, but this opening is very small. The total length of the cave is 216 feet. To give a more exact idea of the site, I prefer to divide the cave in three parts. The first is 90 feet long; it is the only one to offer remarkable details. One can stand upright everywhere in it, the height being 12 to 15 feet. The second part, contiguous with the first, has an opening 3 feet 4 inches high and 4 feet 8 inches wide. It is 45 feet long, and very seldom any higher or wider. . . . The third part . . . is 81 feet long: one can stand upright and it is very damp . . . there are no flints visible, any more than in the second part. . . . Only the first part seems to have been inhabited. . . . At the entrance of the main cave there is a quantity of worked flints. . . . To the right of the entrance, to fill the hollows in the rock is a wall made of small stones, flints, and bones. . . . I would willingly believe that, as the savages of America or Africa make trophies of victims in their tents . . . so the inhabitants of the Pech de l'Azé cave wanted to gather the bones of the animals they killed with the tools which served for the killing. . . . (19)

Some days later, Audierne came back to the cave and added: "To the left, in the entrance, are two blocks fallen from the roof. . . . Under these blocks, 8 feet long, 4 feet wide, and 6 inches thick, one can see a quantity of bones, flints, and black dust. . . ."

So we can note that, in 1828, a layer, black in color, was visible near the entrance and that the "walls" were already present, high over the level of the cave floor; we shall see the meaning of these "walls" later. It should also be noted that the western opening had been considered without interest, an opinion that persisted until our own excavations in 1948.

For a long time little mention, if any, is made of the cave in the literature, although several excavators came to the site.

Among them were Edouard Lartet and Henry Christy, who in 1864 did leave a short description of it:

> The Pey de l'Azé [Pey = Pech] cave . . . is one of the biggest we have visited in Périgord. On some points of its internal perimeter, one can see a breccia with bones going up along the walls, sometimes to a height of 1.50 meters. The same breccia adhere to the ceiling where the vault goes down to that level. One can deduce that, before the cave was excavated, the bone-bearing deposits were very thick at the center. . . . There is no memory in the country round about either of the date of this excavation or why it was done. . . . Today, fragments of limestone are scattered on the ground of the cave, mixed with a kind of soft silt with worked flints. These are more numerous on the sides of the caves, where the silt, stuck against the wall, has been hardened, probably by infiltrations of lime-bearing waters. These flints seem better worked, of more varied shapes, than the ones from Combe-Grenal. We have even found some types, usually quite rare in caves, analogous to the ones [found] in very great numbers in the Moustier cave [probably Mousterian of Acheulean tradition handaxes].
>
> The fauna at Pey de l'Azé is quite rich in mammals; reindeer appears again, with wild ox, mountain goat, red deer (very rare) and horse. . . . (23)

The archaeologists' main interest was focused on the more grouped caves around Les Eyzies, along the Vézère Valley. However, in 1908, Louis Capitan and Denis Peyrony made a tentative excavation at Pech de l'Azé and noted:

> There exists in the cave an archaeological layer, formerly unscientifically excavated. This layer continues on the terrace that lies in front of the cave. There, it is about 1 meter thick but is covered with 3 meters of big blocks of limestone and eboulis* resulting from the collapse of the ceiling, which jutted out in front of the cave as a kind of shelter (17).

Then they gave some details about the discovery of a Neanderthal child skull, details we shall return to later.

Unfortunately no more elaborate report was ever made, since

* *Eboulis:* a French word, for which there seems to be no exact English equivalent, meaning fragment of rock, of any size, fallen from a cliff or the roof of a cave or shelter.

at about the same time Capitan and Peyrony found the first of several Neanderthal skeletons at la Ferrassie. Capitan was a good archaeologist, but in the old-fashioned tradition, and Peyrony was then only a beginner. The material found at that time was all lost, except possibly some handaxes. Long afterwards, some time before his death, D. Peyrony told me that the skeleton had been found just under the present porch of the cave, close to the wall.

Then again a number of people, usually pot-hunters, or "two-legged badgers," as we call them, came and destroyed part of the remaining layers. At that time, no organization existed to protect prehistoric sites. One of them even got himself killed, foolishly trying to dig under the fallen blocks.

In 1929 and 1930, Professor Raymond Vaufrey made his excavations at the site, with more scientific methods and a better understanding of the stratigraphical problems. He established that all the deposits he found in front of the cave belonged to the Mousterian of Acheulean tradition, that is, with handaxes. Two main layers were recognized: a lower one, which contained many handaxes and side scrapers of good workmanship; and an upper one, with less handaxes and scrapers, which were not so well made, and many denticulated tools and notches. Soon after this excavation, Professor Vaufrey became involved in his work in North Africa, for which he is justly famous, and published only a preliminary report of his work in France in 1933 (29).

THE 1948–1953 EXCAVATIONS

In 1947, Professor Vaufrey asked me to undertake the excavations at Pech de l'Azé again, in order to verify some points and to get more material for study. I began in 1948 and continued until 1951 at Pech de l'Azé I (see Fig. 5). It was just after the war, money was scarce, and I worked with the help of one friend, Maurice Bourgon, a very good amateur archaeologist, and a Spanish worker, who broke up the rock. It was not possible to use explosives because of the railway nearby. No elaborate mapping was done (after all, only a small part was to be excavated, and the result would not have had much significance), but the most careful attention was given to the stratigraphy, and all the lithic material, as well as all the bones, were kept for study. Even so, some interesting facts about localization and features were established.

As a result of all the previous excavations, almost nothing was left *in situ* inside the cave. The first third was filled up with old dumps, rich in tools, flakes, and bones, but also rusted sardine

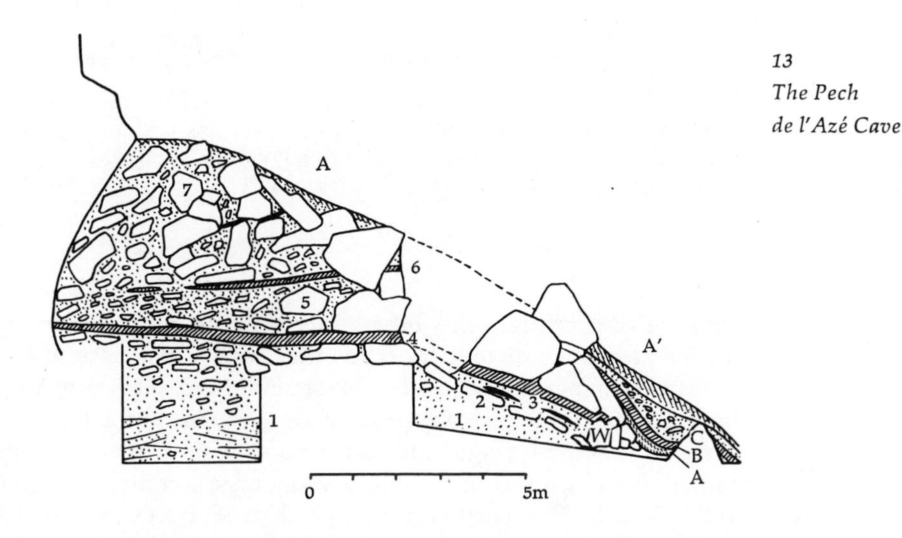

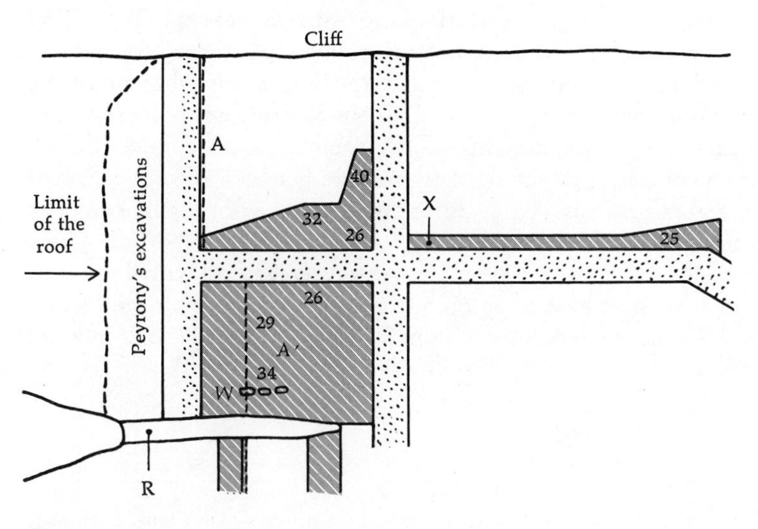

Fig. 5

Top: Section at Pech de l'Azé I, following line AA'. 1, Sands. 2, Pavement(?). 3, Mousterian of Acheulean tradition, type A, lower layer. 4, Mousterian of Acheulean tradition, type A, main layer. 5, 6, 7, Mousterian of Acheulean tradition, type B. A, Mousterian of Acheulean tradition, transitional type. B, C, Mousterian of Acheulean tradition, type B. W, section of the wall.

Bottom: Plan of excavations at Pech de l'Azé I. The dotted surfaces

Fig. 5 (cont.)

indicate excavations by Professor R. Vaufrey; the shadowed surfaces, excavations by F. Bordes.

R, rocky ridge continuing the cave wall. AA′, place of the section. W, wall. X, area where handaxe flakes were very numerous. 32, 40, etc.: percentages of side scrapers in these spots.

cans, broken bottles, and other traces of the meals either of earlier excavators or tourists. The other two-thirds, close to the passage communicating with the second cave, had probably been untouched by excavations: on the ground was a layer of rounded limestone fragments and a number of flint tools with rather shiny surfaces and slightly worn edges, resting on light yellow sand. This part of the cave showed traces of natural action rather than man's. But mixed in with the stones and flints were fragments of wooden shoes, unfossilized sheep bones, and a fifteenth-century coin, probably trodden into the soil by the sheep.

Against the right wall (looking into the cave), about 1.50 meters above the old dumps, were two platforms formed by a hard breccia with bones and flints (Fig. 6). At that point we made the same interpretation of them as Professor Vaufrey, that is, that they were remains of the same deposits found outside. At the entrance to the narrow passage between the caves, small remnants of a somewhat different breccia, with bones, could be seen.

In this narrow passage, we were obliged to crawl on our stomachs or at best to go on all fours before entering the second cave. Contrary to Audierne's description, we found it dry enough,

Fig. 6

Top: Pech de l'Azé I. In front, the old dumps. Over them, breccia I. On the left, the passage, with a wall to contain the old dumps.

Bottom: The entrance to Pech de l'Azé II. On the left, in front, part of the section showing the Riss I and II layers, the eboulis of the beginning of Würm I, and the Würm I deposits. Under the ladder, Riss I and II deposits (sandy), Riss III deposits (eboulis) and Würm I deposits.

In front of the cave, the Riss III eboulis (reddish) has been destroyed and replaced by the yellowish Würm I eboulis.

but the entrance was almost closed by a cone of earth and fallen rock. At the top of the cone, two big blocks framed a narrow aperture, through which a scanty light reached the cave. No clear indication of habitation could be found, except, on the surface, a flat, polished stone which may have been a Neolithic grindstone, and an Upper Paleolithic end scraper (Fig. 23).

We did not try to leave the second cave this first year, being convinced by our predecessors and by what we had just seen that it was probably without interest.

The Excavations at Site I

Practically all the inside of the cave had already been dug up at various times. Outside, there were only the H-shaped trenches made by Professor Vaufrey (Fig. 5). We applied the technique of excavation which was still prevalent at that time. This consisted of noting the horizontal position of the finds only roughly but of taking the utmost care over the stratigraphy. We began in the lower arch of the H, closest to the slope. There, collapsed rocks covered part of the surface only, but a second layer of blocks lay over the lower layer and was in turn partially covered by the upper layers. From bottom to top, the stratigraphy was as follows (Fig. 5):

1. light yellow sand, stratified, sterile, with small eboulis at the top;
2. a pavement, possibly natural, of flat limestone slabs;
3. yellow sand with lenses of ashes, Mousterian of Acheulean tradition;
4. main layer, black ashes, more or less consolidated by calcareous concretions; very rich Mousterian of Acheulean tradition directly covered by big fallen blocks; this layer was limited near the slope by a low wall of stones about 1 foot high; beyond the wall, it did not exist.

For convenience, the upper layers at this spot have been given letters instead of numbers:

A. against the blocks was a layer of reddish dirt, perhaps waterlaid, with rare tools belonging to the Mousterian of Acheulean tradition; upper level 1;

B. blackish layer, ashes mixed with dirt and stones, with numerous tools and bones, Mousterian of Acheulean tradition; upper level 2;

C. yellowish layer, made up chiefly of small roof fragments, with tools and bones; upper level 3;
D. modern soil, with tools in secondary position.

We also made a limited excavation in the upper arch of the H. There, over layer 4, we found:

5. jumble of blocks with yellow sand, a few tools, and bones; upper layer, lower part; Mousterian of Acheulean tradition;
6. blackish layer, with traces of fire among the smaller fragments; upper layer, middle part; Mousterian of Acheulean tradition;
7. jumble of blocks with scattered fire traces, tools, and bones; upper layer, upper part; Mousterian of Acheulean tradition.

Over this, in a part we did not excavate then, were huge blocks of limestone, almost sterile. During the later occupation of the cave, the rock shelter must have been collapsing rapidly. Probably most of the occupation layers were inside and were destroyed before our excavations, perhaps even before Audierne's visit.

So the main result of our work was the confirmation, with more details, of Vaufrey's conclusions, and the discovery that the Mousterians of layer 4 had built a low wall in front of the cave and that the deposits of this layer did not go beyond the wall.

The Discovery and Excavation of Site II

Meanwhile, in 1949, we got through the small opening at the top of the deposits of Pech de l'Azé II. Outside, there was a rather steep slope, partly natural (the outside cone of eboulis) and partly manmade when the railway trench had been dug around the 1860s. This trench had destroyed part of the site, and on the section we could see bones and flints protruding at several levels. It is interesting to note that if in Jouannet's or Audierne's times probably nothing could be seen to indicate the existence of another site, after the 1860s flints and bones were visible on the section. Nobody before us seems to have seen them; there were no traces even of test excavations at Pech de l'Azé II.

We began the excavations at Pech II in 1949, but had only a few days to spend there. We obtained an elementary stratigraphy in the thinnest part of the deposit near the railway trench, running from top to bottom:

1. modern soil, 0.20 m;
2. red sandy earth, with worked flints, rolled and unrolled, 0.35 m;
3. pinkish layer, with rolled flints, and a few unrolled, 0.25 m;
4. stone polygons, a periglacial feature;
5. thick soil; reddish silt, hardened in places, rich in bones (this first year we did not find any implements in the small area excavated, but some of the bones did seem to have been broken intentionally);
6. sands, more or less stratified, with rounded eboulis at the top.

Then the slope was covered with grass and bushes, but lower, inside the railway trench, we could see a breccia with iron nodules, then the bedrock.

We attributed (correctly) the worn aspect of the rolled flints in layers 2 and 3 to cryoturbation, but thought (incorrectly) that they were microlithic tools (see Fig. 19) (11).

In 1950 we began the real excavation. Since the site was untouched, we used a more elaborate technique than before. The site was divided into square meters, a datum line chosen, and the position of all important finds recorded by three-dimensional coordinates, as well as by their situation in relation to the layers (see p. 43). Most of our effort was spent on the lower layers, in the part where they were exposed. However, to facilitate the evacuation of the back dirt, we dug a trench in the double cone at the entrance of the cave and also began another under the shelter. We obtained an elementary stratigraphy of the upper (Mousterian) layers and a good one of the lower layers. From top to bottom this ran as follows:

- S. modern soil;
- 1. sandy yellowish layer, with small angular limestone fragments, 1.20 to 1.40 m; this was then subdivided into 1a, with some flakes and a finely retouched blade fragment, which at the time we thought was possibly Solutrean (in fact, it is Mousterian); 1b—almost sterile; and 1c—with a few tools of Quina-type Mousterian;
- 2. reddish sand with eboulis, 0.50 m;
- 3. sandy clay, reddish, 0.50 (maximum) m, Denticulate Mousterian;
- 4. inside cave II, layer of ashes and stones, 0.20 m, Denticulate Mousterian;
- 5. polygonal soil with, inside the polygons, small rounded pieces of eboulis with small battered flints; underneath was a layer of limestone slabs; 0.70 m;

6. reddish silt, with pseudo-mycelium and big rounded eboulis; just at the top, some flints and horse teeth; lower, rich fauna and flints; horse, oxen, red deer, cave bear, wolf etc.; 0.40 m;
7. yellowish sandy silt, sometimes with a greenish tinge, with flints and the same fauna, plus Merck's rhinoceros; 0.10 to 0.15 m;
8. reddish silt with rounded eboulis, flints, same fauna as in 7; 0.10 m;
9. rounded eboulis in yellowish sands, with flints (one handaxe), same fauna as in 7; 0 to 0.25 m;
10. yellow sands, sterile.

One of the surprising results was the total absence of the Mousterian of Acheulean tradition so plentiful at the other end of the cave (side I). Not a single handaxe had been found in the Mousterian layer, not any of the characteristic flakes. But the top layers were very poor, so the explanation may be that the Mousterians of Acheulean tradition had not lived much on that side and that the Denticulate and Quina Mousterians were restricted to Pech II. The general stratigraphy could have been: Denticulate Mousterian, Mousterian of Acheulean-tradition, Quina Mousterian. The reason for such a distribution, however, was none too clear.

As for the lower layers, layer 9 had yielded a handaxe, which could be Acheulean. Layers 8, 7, and 6 had not given any handaxes but several choppers and chopping tools, and could belong to the handaxeless phylum which comprises the Clactonian.

In 1951 we went on with the excavations, and it soon became clear that the site was more complicated. We published a preliminary note about the lower layers in November 1951, in which we attributed these industries to the "Pre-Mousterian," comparing them to the lower layers of la Micoque, the famous site near les Eyzies (Fig. 2). This note contained one major error about the dating. At that time, Merck's rhinoceros was considered a good indicator of a warm climate preceding the last glaciation. Its presence was even the main reason why the famous site of Krapina, in Yugoslavia, was assigned to the last interglacial. At Pech de l'Azé II, the faunal remains in these lower layers were not "cold," so it was quite natural, at the time, to have supposed that these layers dated from the last interglacial.

The year 1952, which was the year of the main excavation, yielded other data. The stratigraphy of the upper layers became clearer, and it was soon apparent that we had missed a lot in the preceding years. Fortunately, thanks to the method of excavation and the numerous notes taken, it was relatively easy to correct our mistakes. Layer 3 did not contain one layer of Mousterian, but four layers, very close to each other and often in contact. At some places, in 1952, they were well separated by up to 5 centimeters of sterile sands, which helped to correct the mistake. Layer 2, also, was complicated; there were several archaeological levels, very poor but clear. Finally, the following stratigraphy was established, to which the second phase (1967–1969) of excavations added only details and further finds. We had to assign new numbers to the layers (Fig. 7):

S. modern soil, disturbed by rodents;
1. reddish sandy layer, with traces of undeterminable industries; 1 m;
2. yellowish layer, subdivided into:
2A, B, C. yellow sands plus limestone fragments with some Mousterian flints; reindeer present in the fauna; 0.75 m;
2D. sandy clay, with rare limestone fragments; about 50 flints; reindeer; 0.15 m;
2E. numerous limestone fragments with sharp edges; about 50 flints; reindeer abundant; 0.35 m;
2F. smaller fragments in a clay-like sand; only 15 flints; reindeer; 0.30 m;
2G. scattered limestone fragments in sands; 91 flints; reindeer; 0.25 m;
2G′. existing only in places, and found in the second phase of excavations; numerous small pieces of eboulis, often rounded, toward the top; 27 flints; reindeer; 0 to 0.20 m.

Layer 2 was poor in cultural remains, but the simultaneous presence of good Levallois flakes and Quina-type scrapers seemed to indicate an occupation by Ferrassie-type Mousterians.

3. sandy clay, light brown in color, with limestone fragments more or less abundant according to area; 1,261 flints; reindeer in the fauna; Typical Mousterian; 0.30 m;
4. reddish layer, subdivided into:
4A. sandy clay, reddish brown, with rare limestone fragments; 0.20 m. There were also two small Mousterian levels with traces of fire: 4A1, 179 flints, probably Typical Mousterian; and 4A2, 337 flints, probably Typical Mousterian with slight contamination from the following layer;
4B. sandy clay, red brown, very hard when dry, numerous traces of fire; Denticulate Mousterian; reindeer very rare,

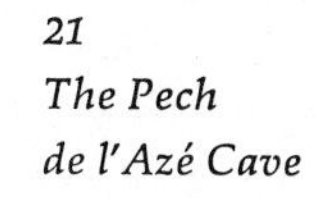

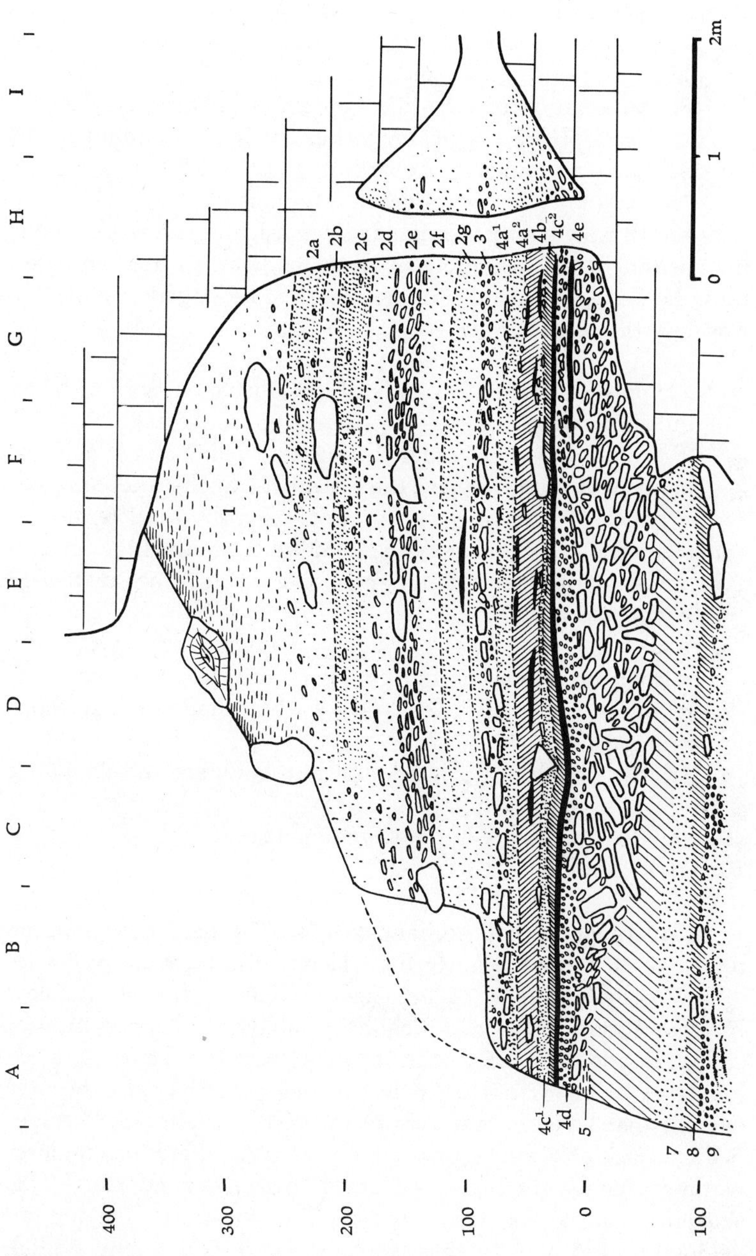

Fig. 7

Schematic section at Pech de l'Azé II. This section is perpendicular to the railway trench.

dominant animal red deer, followed closely by horse; 5,939 flints; 0.10 m;
4C1. sterile in certain areas, 0.05 m maximum; at other places, a mixture of 4B and 4C2;
4C2. alternation of clay-like sands and brown clay, numerous traces of fire; Typical Mousterian; red deer dominant in the fauna, no reindeer; 2,638 flints; 0.05 to 0.10 m.

Layer 4B was mainly rich inside the cave, and layer 4C2 under the shelter. In the preceding excavations, we had not separated these layers. In most places, they were stuck together and 4C1 was only the contact zone.

4D. polygonal soil, garlands of cryoturbation with crushed flints and rounded limestone fragments inside;
4E. found only well under the shelter; yellow, angular fragments with some flints; this is probably the end, uncryoturbated, of layer 4D; and probably Typical Mousterian;

5. a layer of large fragments, 0.50 to 1 m thick in places, almost sterile (some flints and bones), sometimes disturbed by cryoturbation;
6. reddish soil with flints and bones at the top, and toward the bottom; no handaxes, but choppers and chopping tools;
7. yellowish sandy silt; no handaxes, but choppers and chopping tools;
8. reddish silt, thin; no handaxes, but choppers and chopping tools;
9. rounded fragments of limestone; handaxes;
10. sterile sand.

As layers 6, 7, 8 had no handaxes but choppers and chopping tools, we attributed them to the Clactonian after reading Hazledine Warren's article, "The Clacton Flint Industry, A New Interpretation" (31), in which he established the presence of choppers and chopping tools in this industry. Layer 9, with handaxes, was attributed to the Acheulean. It has also become evident that layers 6 and 8 were the result of alteration, under pedogenesis, of 7 and 9 respectively, that is, alteration, under a warmer climate, of deposits formed in colder conditions. This sequence (9 to 6) was very probably Rissian in age. The soil developed on top of 9 (that is, so-called level 8) was interstadial, and the thicker soil developed on top at 7 (that is, 6) probably dated from the last interglacial age. The fact that all through these layers the fauna was a "warm" one did not trouble us too much then. In the "older loesses" of northern France, clearly

Rissian, the fauna found most of the time was *Elephas antiquus* and Merck's rhinoceros rather than the mammoth and the woolly rhinoceros. If the fauna seemed to contradict the geological and pedological facts, it had to be reinterpreted, since, so far as we know, adaptation to cold climates is not evident from bones, and anyway, many animals can live in a nonoptimum environment. So we began to question the value of the so-called warm fauna (without hippopotamus, however) as a sure indicator of a warm climate.

The main reasons why layers 6 and 8 were now considered mere weathering horizons of 7 and 9 respectively, and not true layers, were: (1) the gradual transition from one to the other, with a gradient in clay content and a gradual change of color; and (2) the existence of a pedological B horizon of accumulated carbonates under the guise of pseudo-mycelium found at the top of 6 and in 8. When weathering occurs, there is a leaching of the lime from the top of the layer and a concentration toward the bottom, producing different types of concretions. In the same way, this part of the layer is enriched in clay and small particles; the top, impoverished in small elements, becomes an easy prey to erosion and is usually not preserved in fossil soils.

The Riddle of the Breccias

In 1951, on a rainy day when it was not possible to work outside, we did some excavations in the "suspended breccias" stuck against the wall inside cave I (Fig. 5). We were rather surprised not to find any handaxes, since we thought then that these breccias were the remains of the Mousterian of Acheulean tradition layers inside the cave. Also the flints were slightly shiny and smooth, as happens under very damp conditions when sand is driven over them by small rivulets. At that point we had no clear idea of what kind of industries existed in cave II, and so we attributed the absence of handaxes to the small area excavated in the breccias and to possible specialization: handaxes would have been used only in the front part of the cave. However, the style of these flints was very different from that of Mousterian of Acheulean tradition, and the two assemblages large enough to be statistically analyzed looked suspiciously like Typical Mousterian and Denticulate Mousterian.

So, when we discovered what types of industries did exist in cave II, it looked very much as if we had, in these breccias, not a continuation of the Mousterian of Acheulean tradition deposits, but rather a part of the stratigraphy of cave II, and more precisely a part of layer 4, with Typical and Denticulate Mousterians. And it dawned on us that these breccias were probably the remnants of Audierne's "walls," partially excavated by someone (the mine holes made for the explosives used to break the breccias are still visible).

It had been necessary, in excavating side II, to pass the back dirt all through the cave, since the presence of the railway made it impossible to dump it on the slope. For this reason, we had been obliged to dig a trench through the old dumps in cave I (we found isolated blocks of the breccia in them) and through the narrow passage between the two caves. In the passage, we found a great many cave bear teeth and bones in the sand, most of them unfortunately broken by the rabbits that made warrens in this protected spot. In the rear part of cave I, between the old dumps and the passage, we found, all mixed together, Quina-type scrapers (unknown in the Mousterian of Acheulean tradition, with one or two exceptions, differently patinated), and other worn flints of the type found in Pech II in layers 6 to 9 (no handaxes, however), basalt choppers and chopping tools, denticulates galore, and a very few handaxes of the Mousterian of Acheulean tradition, these last in a much fresher condition.

All this added up to a riddle: here we had a cave open at both ends, with Mousterian of Acheulean tradition only at one end, and at the other end Acheulean, Clactonian (or so we believed at that time), Typical Mousterian, Denticulate Mousterian, Typical Mousterian again, then Quina-Ferrassie Mousterian. And part of this sequence was found in the breccias stuck against the wall of cave I, while Quina-type tools were found, mixed with older stuff, among the rounded debris in the inner part of cave I.

Only one explanation seemed possible: during Riss times, first Acheuleans, then Clactonians inhabited both caves, and in both caves deposits accumulated and weathered at the top during the last interglacial. Then, also in both caves, came Typical Mousterian, Denticulate Mousterian, and Typical Mousterian again; then Quina-Ferrassie Mousterians lived there during the first part of the Würm, and Würm I deposits accumulated. At the end of Würm I, cave II was almost completely filled at the entrance, and a double cone extended in front of the cave as well as inside. Cave I was probably also almost filled, and the deposits extended far inside, at least to the present position of the breccias (Fig. 8).

We know, from studies of the loesses of northern France, that during the Würm I/Würm II interstadial heavy erosions took place, probably under very damp conditions. At Pech de l'Azé, water came into the cave by lateral ramifications and/or through

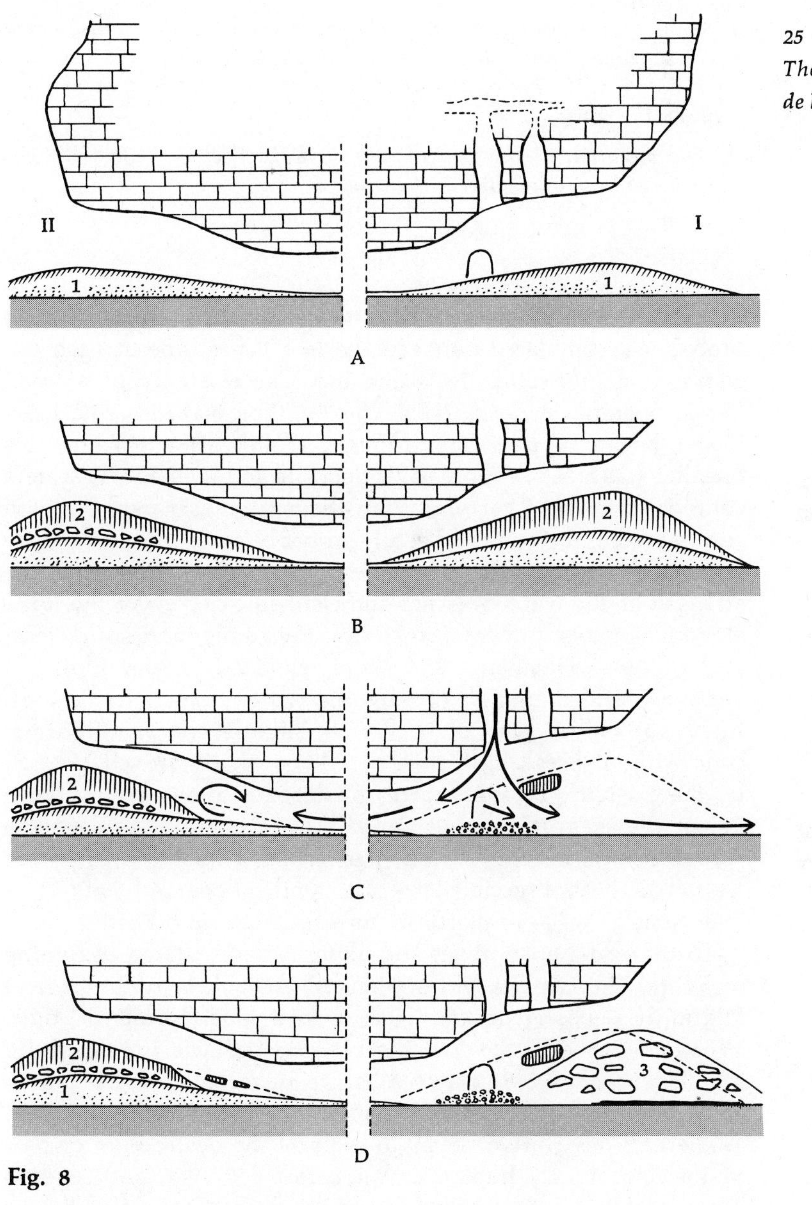

Fig. 8

History of the Pech de l'Azé Cave.

A. The situation after the Riss-Würm interglacial. A soil developed on top of the Rissian layers (1).

B. The situation at the end of Würm I. Würm I sediments (2) have been deposited over the last interglacial soil.

C. Water coming in during the Würm I/Würm II interstadial. De-

Fig. 8 (cont.)
struction of the sediments at Pech I, leaving only the consolidated breccias against the wall in Pech I and some traces in Pech II, and some remains of layers, disturbed, near the communication passage between the caves.

D. The situation at the end of Würm II. In Pech I, Würm II sediments have been deposited in the cave (3).

the ceiling, where holes can now be seen in the rock. This water probably accumulated between the two cones, one at each entrance, and infiltrated the sediments. The existence of a small lateral opening in Pech I (Fig. 3) doubtless played a role here. Water began to percolate through it, and after a while this opening was freed of most of its deposits and the pent-up waters could rush outside, carrying with them the closer deposits, and so weakening the cone of Pech I, probably not strong enough to withstand the pressure any longer. In the rear of the cave, the strength of the water was not sufficient to carry away the limestone fragments and the flints, only the sands; and so, at least part of these fragments and flints remained on the spot, but became rounded or polished by the rushing sand. In Pech II, there was only a kind of backwash, slightly eroding the inner cone without breaking through it. This can be proved (Fig. 8) by the existence of some traces of deposits above ground level before the excavations. But, on side I, the washing-out process was complete, except for that part which (situated against the wall) had been brecciated by stalagmitization, and which remained as a "wall" or platform, now well up into the air.

To that side, open again and almost empty, at the beginning of Würm II came the Mousterians of Acheulean tradition, and Würm II deposits accumulated over a long period of time. Whether they ever covered the breccias again is impossible to say. The breccias were exposed before *archaeological* excavation, after Audierne, but perhaps not before excavation of some kind. Audierne's description seems to indicate the destruction of part of the deposits by human action before his visit, and so does Lartet's. We shall find a similar case at Combe-Grenal.

This then was the story when, in early September 1953, we gave up the excavations at Pech de l'Azé to undertake those at Combe-Grenal. The general explanation seemed to hold, but there were a lot of loose ends. The correlation between the two caves had been established only on archaeological grounds, and could be disputed. Also, we knew that the Riss glaciation consists of at least three phases, and we had found only two at Pech II. Which of them was not present, and why? The dating of the Mousterians on side II as Würm I, and on side I as Würm II was also disputable. Why could not the big erosion date from within Würm I, for instance? The only reason we dated it from an interstadial was that, in northern France, 400 miles away, the erosion was so dated. There was no stratigraphical proof that an interstadial had occurred between the Quina-Ferrassie Mousterians on one side and the Mousterian of Acheulean tradition on the other. But at that time we felt that these questions could not be solved without more powerful methods than the ones at our disposal.

2. *New Methods*

I worked at Combe-Grenal for 13 years (September 1953 to September 1965) and at Roc de Combe for one year (1966) before I was able to come back to Pech de l'Azé. Meanwhile, there had been many changes. Instead of a graduate student, then an almost isolated researcher at the National Center for Scientific Research, I was now a professor at the University of Bordeaux, with a laboratory, assistants, students, and a whole array of technical approaches to the problem. And I had accumulated 14 years of experience!

When we undertook the excavations at Pech de l'Azé II again, at the beginning of July 1967, natural erosion, after 14 years, of the more or less well protected sections, plus the vandalism of "two-legged badgers," made it a difficult task. For instance, some unknown fool had very carefully and thoroughly pulled off the iron spikes which marked our datum lines and planted them again at random, had scratched away the paint spots, and so on. So the excavation grid could only approximate (give or take 10 to 20 centimeters) the original one. Part of some layers had been "excavated" by the "oven technique" so dear to pot-hunters everywhere. However, as a whole, not too much damage had been done.

Our aims were to enlarge the excavations and extend them into the cave, to get more material from the not-too-rich layers, and to do all the sedimentary and pollen analysis, the paleontological studies, and so on, which had been at best sketched during the first excavations.

Under my general direction, the permanent team consisted of François Prat, Ph.D., assistant professor at the university, a paleontologist; Miss Marie-Madeleine Paquereau, Ph.D., a palynologist; and Henri Laville, a graduate student working for his Ph.D. whose subject was caves and shelter deposits. All three belong to the Laboratory of Pleistocene Geology and Prehistory at the University of Bordeaux and I shall borrow much from their work in what follows.

SEDIMENTOLOGICAL METHODS

The sedimentology of caves and shelter deposits was first systematically studied by the German Robert Lais (1941), then by several others, among whom the German Elisabeth Schmid (1958), the Hungarian Lazlo Vértes (1959), and the Frenchman Eugène Bonifay (1955–1962) are the most important. Laville's work stems from these previous works, with improvements and necessary adaptations.

The principal aim is to examine the mechanisms of sedimentation and of climatic variations. In our latitudes, temperate and damp conditions cause a chemical desagregation of limestone and sandstone, and cold and drier conditions cause a physical erosion. By frost action, parts of the wall or the roof are detached from the rock and fall to the ground. This accumulation is, of course, faster in the open shelters than in the almost closed caves, where it occurs mainly at the entrance, giving the double cone of deposits (see Fig. 8).

Incidentally, this climatic action is the origin of most of the shelters and also enlarges the entrances of caves. Figure 9 shows a general history of the formation and decay of a shelter under glacial conditions.

The cold periods produce what is called cryoclastic eboulis, or congelifracts, with angular edges; the more temperate periods, weathered and more or less rounded fragments.

Of course, adaptations have to be made in accordance with the sites. The exposition of the entrance, the nature of the rock, the place where the samples were taken, the proximity of a big palaeolithic fireplace, are all examples of factors which have to be taken into account. If the cave has been closed with skins or wooden logs, eolian deposits may exist outside, but not inside, and so on.

The first essential for sedimentary analysis is correct collection of the samples. One of the characteristics of shelter deposits is their heterogeneity. In the same layer, side by side, may be large fragments, smaller ones, sands, and silts. The presence of worked flint and faunal remains augments this heterogeneity. Then there is often an alternation of thin layers, some with a coarse texture, others with finer-grained sediments, and layers should no more be mixed in sedimentary analysis than in archaeological analysis. For study of horizontal variations one should, whenever possible, take samples from different places in the same layer. One of the characteristics of our ways is that the sedimentologist takes part in the excavation and stays several days, so that he has a good firsthand knowledge of the problems

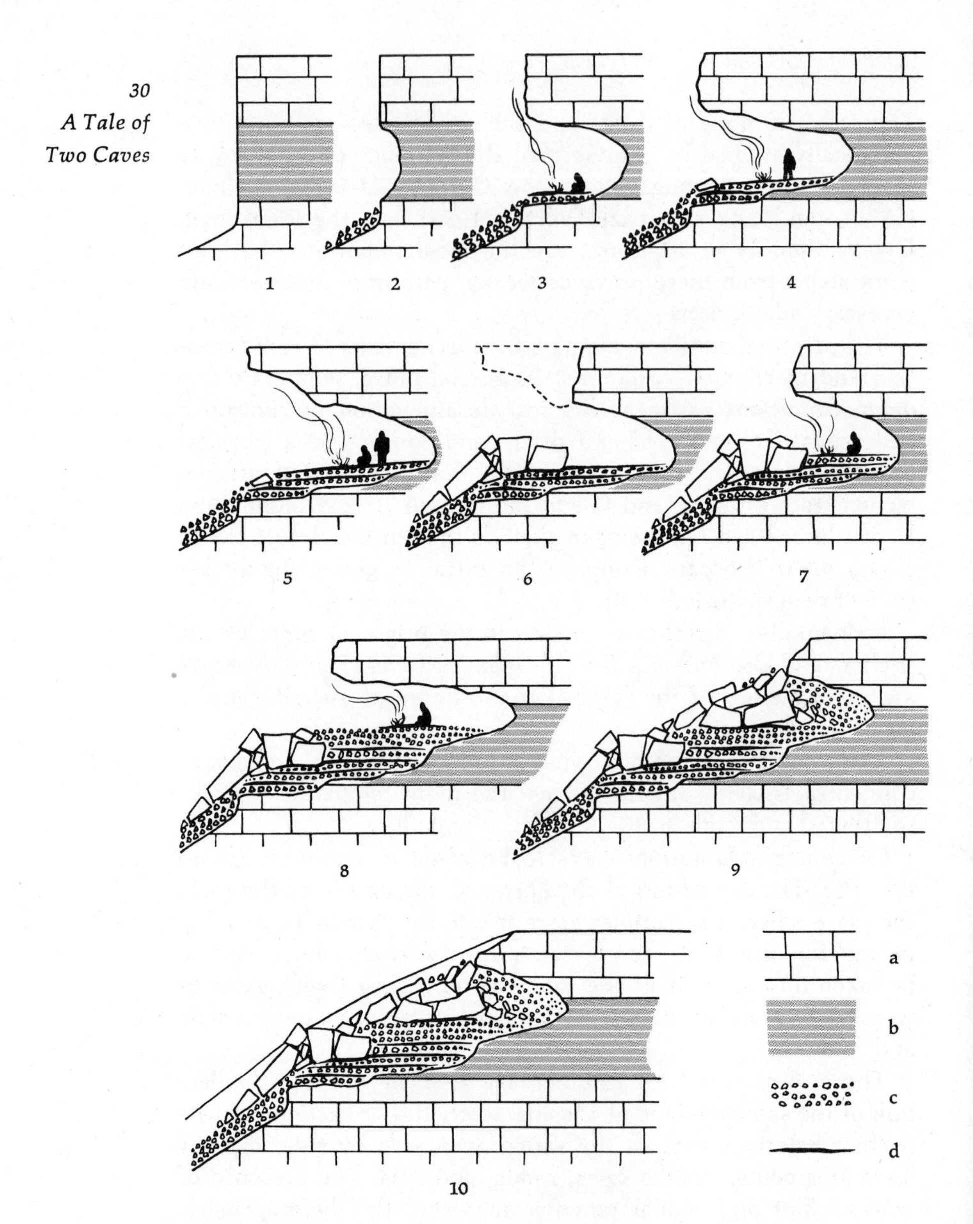

Fig. 9

Formation and evolution of a rock shelter.

a. Compact limestone. b. Frost-sensitive limestone. c. Eboulis (congelifracts). d. Archaeological layers.

Under frost action, an embryonic shelter is formed (1 and 2), which becomes deep enough to be inhabited (3). The frost action going on, the shelter becomes deeper and deeper, but the parts protected by

Fig. 9 (cont.)
the eboulis are not attacked any more. Hence, the formation of steps in the bedrock (4 and 5). Then there is a first collapse of part of the roof (6), with an interruption of inhabitation by man. Man comes back (7), and the shelter goes on deepening (8) until there is a general collapse of the roof (9). Colluvia cover the whole (10), as in the case of la Ferrassie, la Micoque, etc. In other cases, part of the shelter remains open.

There are some periods where frost action is more active than others, and this also contributes to the formation of the steps in the bedrock of the shelter. Note that the older the layers, the more they are situated to the front of the present shelter.

of the dig. We do not believe in the "experts" studying in a laboratory samples of a site they have never or barely seen.

The first thing the sedimentologist does at the site is to discuss the stratigraphy with the head of the excavation and make sure he understands what is going on. Then he carries out a descriptive study of each of the layers: thickness, and its variations; color; texture; possible cryoturbations; and so on. The samples are then taken in stratigraphical continuity: that is, not only one for each layer, but several if the layer is thick, the bottom of one sample being the top of the following one. Samples for pollens are taken, if possible at the same time, by the pollen analyst in contiguity with the sedimentology samples, and with at least a one-to-one correspondence. Often several palynology samples correspond to one sedimentology sample.

If the sediment is not too rich in eboulis, 200 or 300 grams are enough for sedimentology. If there are many eboulis, it is necessary to take from 5 to 15 kilograms, to be sure that enough fine-grained sediment will be present. Once the samples have been taken, the long, tedious, but fruitful laboratory work can begin.

Global Granulometry

First a "global granulometry" is done. All the blocks over 100 millimeters are taken out and held separately. The sediment is dried up, and three sieves are used, 10-mm, 5-mm, and 2-mm mesh, respectively. Four fractions are thus separated:

between 100 and 10 mm: blocks (group I)
between 10 and 5 mm: granules (group II)
between 5 and 2 mm: gravels (group III)
below 2 mm: sands, silts, and clays (group IV)

Each class is then cleaned of any foreign element brought by man, such as flint flakes and chips, pebbles, bones and bone splinters. Then a stratigraphical diagram is built by cumulating, for each level, the relative percentage in weight of the four granulometric classes (Fig. 10.1). This diagram shows the intensity of cryoclastism (frost action) and its evolution in the stratigraphical series. The more there are of groups I and II, the more frost action was important.

Granulometry of Blocks

The elements between 100 and 10 millimeters are sorted into nine granulometric classes (100 to 90, 90 to 80 . . . 20 to 10 mm). Each class is weighed and a diagram built up by cumulating the percentages in weight of each class for all the layers (Fig. 10.2). This shows the modalities of thermoclastism. If there are a lot of large elements, this indicates an alternation of great amplitude, in value and length, of freezing and thawing. If there are few large elements, this probably reflects a seasonal variation of these phenomena.

Variations of Percentage of "Frost Slabs" in Group I

"Frost slabs" are flat debris, with angular outlines—if they have not been rounded by later natural action. One of the faces is fresh, the one that was inside the rock before the slab was detached and fell; the other face often shows some weathering. This study is done on a given granulometric class found all through the stratigraphy (for instance, 20 to 60 mm), and the percentage in weight is calculated. This enables us to calculate the modalities of frost actions more precisely by working out the difference in percentage between these slabs and the ordinary polyhedral eboulis (Fig. 10.3).

Another study bears on "frost-cracked" blocks. These have been cracked by frost action after their deposition in the sediment. Their fragments are found together, either because the cracking was not complete or because they have been secondarily cemented by lime concretions (Fig. 11). Their percentage is calculated in relation to the totality of the blocks, and their significance is that they indicate frost action during or after deposition (Fig. 10.4).

Quite frequently in cryoturbated levels one finds a different kind of block, the so-called fissured block, which has superficial cracks, often in a sort of network. These seem to indicate unstable climates.

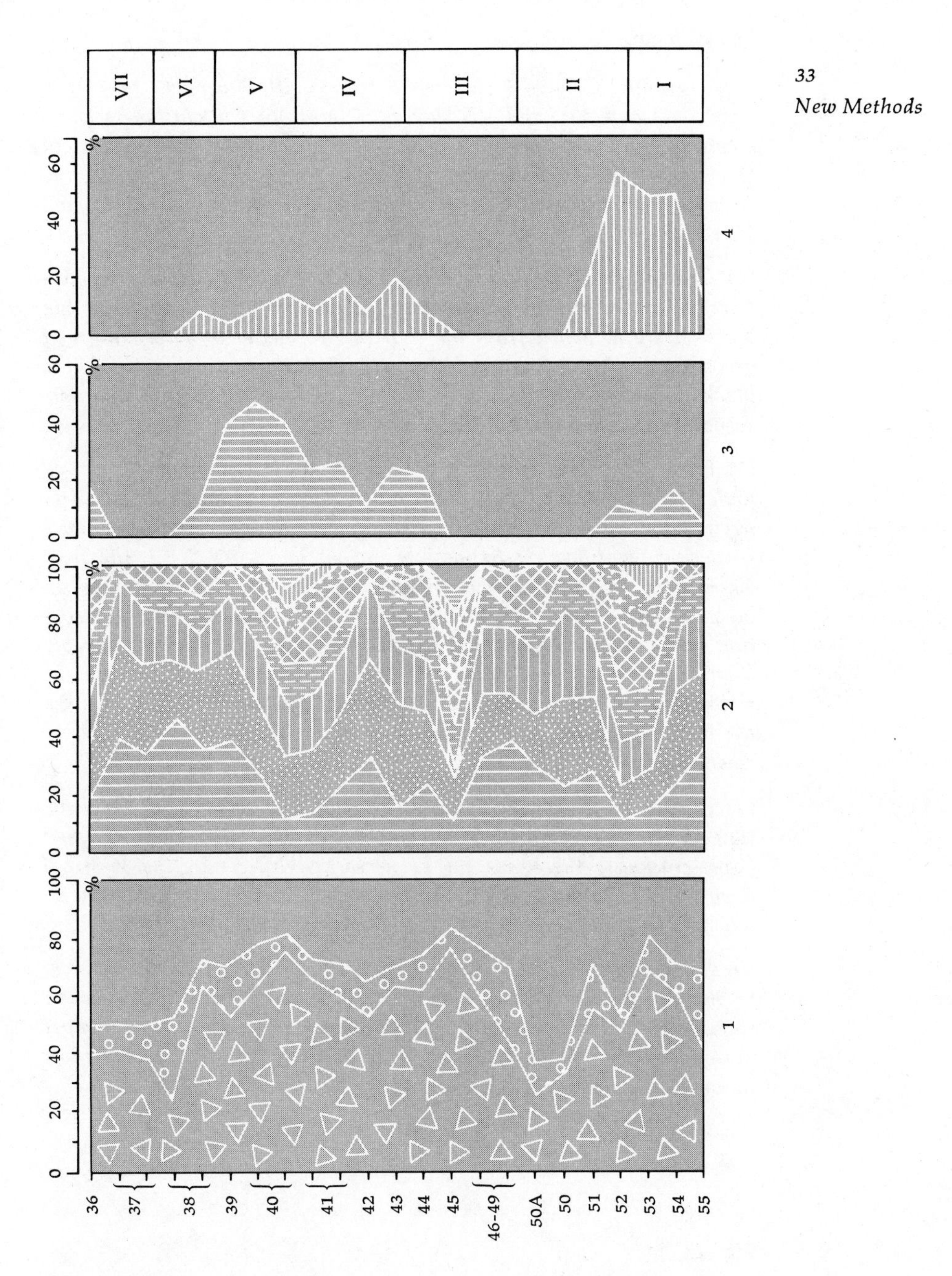

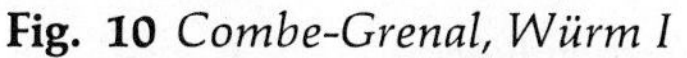

Fig. 10 *Combe-Grenal, Würm I*

(1) Global granulometry, in three-dimensional categories, gives information on the composition of the sediments and their evolution in the section. This first analysis permits us to divide the layers into

Fig. 10 (cont.)

seven groups. Groups I (layers 55 to 53), III (layers 49 to 44), V (layers 40 and 39), and VII (layers 37 and 36) contain layers characterized by a high proportion of coarse elements. They correspond to periods when frost action was more important than during the formation of groups II (layers 52 to 50A), IV (layers 45 to 41), and VI (layer 38).

(2) Granulometric analysis of blocks between 100 and 10 mm precises the modalities of frost action. In groups I, III, V, and VII, big blocks were detached from the roof of the shelter by alternations of great amplitude of frost and thawing. In groups II, IV, and VI, the blocks, smaller, were detached during less cold conditions and correspond rather to seasonal frost action.

(3) Frost slabs are usually more numerous in sediments deposited under cold conditions. But their morphology is often modified, after deposition, by alteration or cryoturbation phenomena, sometimes rendering identification impossible. So, in layers 55 to 53, the frost slabs seem rare in a sediment which, however, is rich in eboulis. In layers 51 to 46, all these elements have been rounded by cryoturbation and no typical frost slab subsists. In layers 37 and 36—deeply weathered during the Würm I/Würm II interstadial—almost all the slabs have been disfigured by dissolution. It is only in layer 40 that the high percentage of slabs seems to match the intensity of thermoclastism.

(4) Eboulis once detached, are still under the influence of frost action. When this action is strong, the blocks split into large fragments (frost-cracked); there are no frost-cracked blocks in the Würm I deposits of Combe-Grenal. On the other hand, under seasonal or day-to-day phenomena, the effect of frost action is less strong.

This is the case in the Würm I deposits at Combe-Grenal, where it is mainly in the layers underlying the less thermoclastic levels (layers 52 to 54, 44, 39), or contemporary to less cold but damper episodes (layers 42 and 43), that the fissurated blocks are the most numerous.

Fig. 11

Top: frost-cracked block. Middle: fissurated block.

Bottom: effects of dissolution on congelifracts.

The maximum dimension of the biggest block is 10 cm (4 ins).

Calculating the Bluntness Index

The blunting of the blocks is the result either of dissolution or of cryoturbation phenomena. Dissolution attacks first the angle of the blocks; cryoturbation blunts them by mechanical action (crushing of the edges, polishing by rotation and friction). The dissolution effect is strong when, after a cold period, a damp one follows, with plenty of water percolating through the sediments. Blocks of a given class (for instance, 20 to 60 mm) are studied and classified in four categories:

1. not blunted;
2. a little blunted;
3. blunted;
4. strongly blunted.

The percentage in weight of all categories is then calculated, and as it is the "blunt" characteristic which is sought, to reinforce this the percentage of category 1 is multiplied by 0 (that is, not taken into account), of category 2 by ⅓, of category 3 by ½, and of category 4 by 1. The total of these percentages, multiplied in such a way, is the index of bluntness (Fig. 12.1).

Measuring Porosity

As we have seen, two very different actions (dissolution and cryoturbation) can blunt the eboulis, and it is important to be able to distinguish between them. For this, porosity is evaluated, by the capacity of absorption of water in relation to 100 grams of dry sediment. The blocks which have been submitted to

Fig. 12 *Combe-Grenal, Würm I*

(1) The values of porosity and bluntness indexes, when linked to alteration by dissolution of carbonates, show a parallel variation. On the other hand, a low value of porosity can sometimes correspond to a high degree of bluntness; in that case, the bluntness of the eboulis is a result of cryoturbation phenomena.

At Combe-Grenal, in the Würm I, it should be noted that the maximal values of porosity correspond to the less thermoclastic zones (group II, IV, IV). Each climatic amelioration has been accompanied by an increase in dampness, and the curve of variation of the bluntness index adds more precision. If one admits that the value of the bluntness index in layer 40, for instance, corresponds to an almost nonexistent alteration, then we can see a clear increase

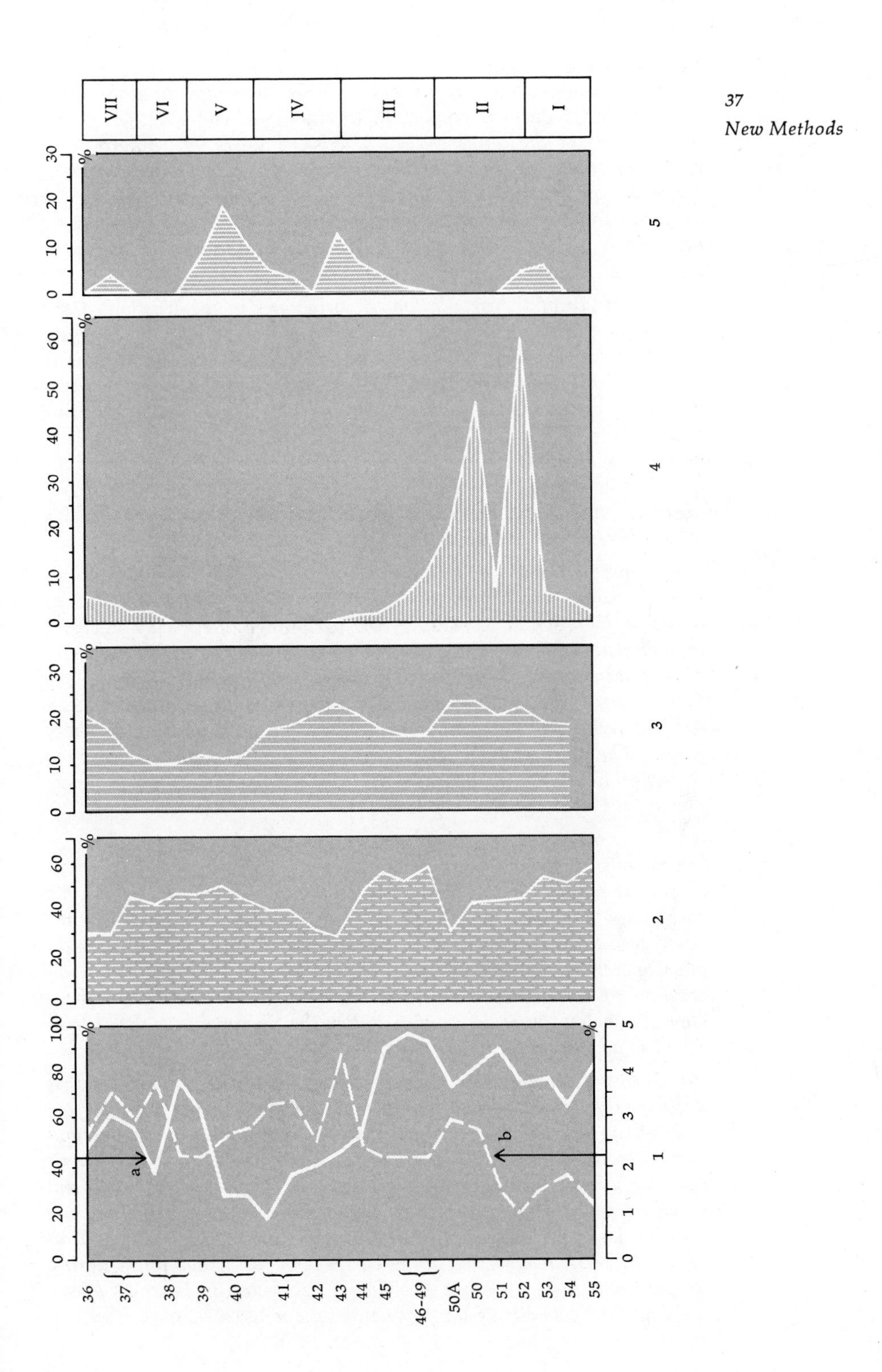
VII
VI
V
IV
III
II
I
5
4
3
2
1
a
b
36
37
38
39
40
41
42
43
44
45
46–49
50A
50
51
52
53
54
55

Fig. 12 (cont.)

in this value in the layers with strong porosity. In that case, porosity and bluntness are linked to alteration. However, the maximum values of the bluntness index exist in layers where porosity is relatively small. In that case, the increase in the bluntness index is the result of cryoturbation phenomena, which did occur during the oncoming of more temperate periods, and manifested themselves by movements of the sediments and blunting of the eboulis deposited during the preceding episodes (layers 55 to 52, 49 to 45, 39).

(2) The study of the proportion of carbonates in the fraction of the sediments inferior in size to 2 mm precises the intensity of weathering phenomena, which, during damp periods, dissolves the carbonate portion of the sediments. Here, the comparison of the curve of variation of porosity with that of the proportion of carbonates confirms that groups II and IV have been deposited under a very damp climate. The dampness contemporary with the deposition of group VI was not so strong. The important decalcification of the sediments in layers 37 and 36 is the result of the very damp conditions of the Würm I/Würm II interstadial.

(3) During a damp period, the alteration of the eboulis and calcareous sands resulted in the liberation of the clay material contained in the rock. The value of the percentage of the clay fractions in sediments of less than 2 mm gives an indication of the intensity of the weathering. Here, three peaks appear on the diagram: the first (in layers 52 to 50A) and second (corresponding to layers 43 to 41) are linked to the dampness at the time of deposition of these layers. The third (which appears in layers 37 and 36) comes from the migration of clay liberated by the weathering of these layers during the interstadial.

(4) The carbonates, dissolved, migrate in depth and crystallize as concretions. The presence of these concretions in layer 52 is linked to the dampness in the deposition of layers 51, 50, and 50A; the carbonates dissolved during the formation of layers 43 and 42 have percolated through layers 44 to 49 and concentrated mainly in layer 50. Concretions present in layers 38, 37, and 36 come from the dissolution of carbonates during the interstadial. The absence of these elements below layer 38 confirms that the dampness of the corresponding period was not very marked.

(5) During the warmer and damp periods, stalactites (P concretions) developed on the walls and ceiling of the shelter. They broke free and fell during the cold climatic periods immediately thereafter and their presence in layers 55 and 54 proves that there was some dampness before the formation of these layers. The P concretions found in layers 49 to 43 had been formed on the walls of the shelter during the damp period corresponding to the deposition of layers 52 to 50A. The ones in layers 41 to 39 correspond to the dampness during deposition of layers 43 and 42. Then those found in layer 37 developed on the walls during the formation of layer 38.

humidity action after their deposition will be more or less porous according to the intensity of the alteration.

One or two kilograms of blocks are dried and weighed, in a given class (20 to 60 mm, for instance). Then they are put to soak in water for at least four hours. They are dried lightly with a cloth and weighed again. The difference in weight is calculated in percentage against the weight of the dry sample.

The bluntness and porosity indexes vary in a parallel way, *except* when bluntness is a consequence of cryoturbation (Fig. 12.1). Sometimes, bluntness can result from trampling of the eboulis by man.

Study of the Concretions

The carbonates, leached out of the accumulated fragments, migrate toward the lower layers and recrystallize as illuvial concretions. Usually, they are aggregations of small blocks or sand grains, and work is done separately on blocks, granules, and gravel fractions. This type of concretions is known as concretions "S" (short for soil), since they are formed inside the soil. But there is another type of concretions, concretions "P" (short for *paroi,* the French for "wall"), which are formed on the walls and roof of the cave or shelter. This does not have the same significance as concretions S.

Concretions S are the result of leaching of the lime from the top of the deposits, under damp conditions *after* the deposition of the layer. Concretions P, on the other hand, are formed on the walls under damp conditions, and are frost-detached from the walls. They correspond to a damp period *before* the formation of the layer. A special example of concretions P are stalactites or fragments of stalactites (Figs. 12.4, 12.5).

Analysis of Fractions under 2 mm

Two methods are employed here. First, calcimetry, in which 100 grams of homogenized sediment are treated with hydrochloric acid. The insoluble part is washed to a neutral pH to take away the excess acid and the calcium chloride, then dried and weighed. The difference from the original weight represents the carbonates, weathered layers being less rich in carbonates than the others (Fig. 12.2). Second, granulometry, in which 100 grams are cleaned of organic material by attack with hydrogen peroxide. The granulometry of the fraction between 2 and 0.05 millimeters is done on a column of 12 sieves. For the fraction below 0.05 millimeters, it is done by densimetry, by application of Stoke's Law.

Cumulative graphs are then drawn, and these yield information about the mode of deposition of the fine fractions of the sediments. A global diagram is also built up by cumulating the percentages of three fractions: 2 mm to 0.05 mm = sands; 0.05 mm to 0.002 mm = silts; and below 0.002 mm = clays. The respective percentages of these three fractions give us information on the illuviation and the degree of pedogenesis in the layer.

The same analysis is also made on decalcified sediments, to establish the influence of the presence or absence of carbonates on the cumulative graph. If the layer analyzed was weathered, the differences between the graphs are small. Further, it is sometimes possible to detect the presence of colloidal carbonates in the "clay" fraction.

Morphoscopy of Elements Around 0.8 Millimeters

These elements are mainly grains of sand, usually quartz grains. If they are rounded and dull, they are probably wind-borne grains, indicating strong eolian phenomena. Most of the time, this means a semi-desert climate, either hot or cold.

Measurement of the pH and Its Variations

The pH is a measure of acidity. It corresponds to the cologarithm of the concentration in H^+ ions in a free state in the soil. Twenty grams of sediment, all of less than 2 millimeters, are put into 50 grams of demineralized water for 24 hours. Then the pH is taken with an electrical pH meter. At the same time another sample has been prepared, but in a normal solution of potassium chloride (KCl) instead of water, and the pH is again measured. The difference between the pH in water and the pH in KCl is called ΔpH. The more weathered the sediment, the larger the ΔpH, probably because of the alteration of the clay minerals. This method is quite sensitive and permits the detection of weak weathering in the layers.

X-ray Diffractometry

A study of the clay components of sediments can be done by X-ray diffractometry. Both the nature of the clay minerals and their state of preservation can be identified, giving information on the origin of the sediment and its degree of alteration.

Chemical Analysis

Some promising results have been obtained by the study of migration in the layers of several elements (calcium, sodium, iron, manganese, etc.), but these methods have not yet been applied to Combe-Grenal or Pech de l'Azé.

PALYNOLOGICAL METHODS

The principle of palynology is well known. Pollen grains are almost indestructible and easily wind-borne, so they can scatter over great distances. When they fall to the ground, they are incorporated into the sediments in proportion to their abundance. As plants are usually very sensitive to climatic variations, the determination of fossil pollens, and their relative numbers, can give excellent information on the climate prevalent in a given region during the time of deposition of sediments. They sometimes can also give us a good indication as to the date at which the sediments were deposited, since some archaic types disappeared during the Pleistocene. However, things are not quite as easy in practice as in theory. Some sediments do not preserve pollens: one of the most heartbreaking examples is the lower loam of Swanscombe (England), in which at most one or two grains of pine pollens have been found, which is not very helpful. Furthermore, there may be contamination by more recent pollens, brought into the layers by animal burrowing (by rodents, worms, ants, etc.), or, in the case of coarse sediments, by the holes between the fragments. If these pollens are much younger than the original ones, they can be recognized. But if the contamination is almost the same age as the deposition of the sediment, one can get into trouble. Suppose that the sediment were deposited some 40,000 years ago, under very cold Würm II conditions, and that, during the warmer conditions of the Würm II/III interstadial (about 35,000 years ago), there had been contamination. The difference in preservation of the pollens is not large enough to point up the distinction between the original pollens and the contaminants. This is why we always check pollens against the results of sedimentary analysis.

The utmost care has to be taken when collecting samples for pollen analysis. As in sedimentology, we try to have as complete a sequence as possible. Samples are taken every 5 centimeters, sometimes even more often, after cleaning the section on several centimeters of thickness to avoid modern contaminations. The tools used have to be completely clean. The samples are immediately put into plastic bags, hermetically sealed, and labeled with all the necessary information. When the sediments are fine-grained, a small metallic tube is pushed into them, then closed at both ends.

Back at the laboratory, the long preparation begins. The sediments are washed off stones and pebbles and collected. Carbonates are eliminated by hydrochloric acid. What remains is dried, and put on a 200-micron mesh sieve. The part that passes through is put in Thoulet's solution (a mixture of potassium and cadmium iodides, which has a density of about 2) and centrifuged at a slow speed (1,000 to 1,500 rpm) for 30 minutes. Siliceous material (which has a density of about 2.5) is separated from the pollens and spores, which are lighter. The part in suspension is centrifuged again at a higher speed (3,500 rpm) for 20 minutes, then filtered. What remains on the filter is treated with hydrofluoric acid, taking special precautions (HF is *very* dangerous) to destroy what could remain of fine siliceous material mixed with the spores and pollens. The remnant is then washed until it is neutral. The material thus isolated is kept in distilled water, then mounted on hermetically sealed slides and examined under the microscope. This method seems to be one of the best, since it enables us to find pollens where other methods often fail.

When the pollens and spores have been determined and counted, pollen diagrams are drawn. Here one must take into account not only the presence or absence of certain types of plants, but also their relative proportions (see Fig. 13).

ARCHAEOLOGICAL METHODS

Archaeological methods can be divided in two sections: the excavation itself and the study of the archaeological material, or assemblage.

Methods of Excavating

Excavation methods can and must vary according to site, depending on the specific problems and difficulties encountered. The old system of excavating a trench in the middle of a site should, of course, be discarded, even for a first test. The horizontal excavation, uncovering layer after layer on a wide surface, is theoretically the best and should be used whenever possible. It is the one that can show structures and patterns on the living floors to the best advantage. The catchwords here are "whenever possible." In a cave site, this is very seldom the case, at least in the Dordogne, where very often the occupation was more or less continuous, and there is seldom any sterile layer to separate occupation layers. These can be differentiated, besides their archaeological content, only by changes in the structure and/or color of the sediments, and sometimes such changes may be very slight. In addition, they can be as marked horizontally as vertically, so that a reddish layer on top of a yellow one in part of the site may change, gradually, to a yellow layer on top of a reddish one some feet farther on. Horizontal excavation on wide surfaces in such a context leads most of the time to disaster, even if the excavators are careful and experienced. One goes from one layer to the next without knowing it and, after a while, is completely lost. Even in a case where the excavated layer rests on top of a sterile layer, it is necessary to be very wary. If the layer is thick, it is probably made of lenses, since debris does not accumulate at the same rate everywhere, so that a structure in one part of the site may not be contemporary with another some meters away, even if they seem to be on the same level. And although the layer may be thin, there is no guarantee that the sterile layer which separated layer A from layer B here will be present everywhere in the site. Rich caves are not the best places for palethnological research: they can give a good stratigraphical and typological sequence, but as far as palethnology is concerned, not-too-rich sites (either in caves or shelters) or open-air sites, where the surface was not usually limited, are better.

Thus, most of the time the search for palethnological data in rich caves, if not quite hopeless, is certainly very difficult and will often give dubious answers. One should concentrate on as good a stratigraphy as possible, since stratigraphy is in any case the basis of all research, and without it other work rests on shaky foundations. For this purpose, semivertical (or if you prefer, semihorizontal) excavations are better than horizontal ones.

This method consists of excavating moderately wide surfaces, no more than 4 square meters at a time in each layer. Sometimes, when the stratigraphy is difficult, no more than 1 square meter; in Laugerie-Haute East, for the Magdalenian layers, we took only one-fifth of a square meter at a time. So, the sections will always be close to the part excavated, and a good stratigraphical control is possible. When a structure is discovered, the excavation is enlarged, of course. As for the horizontal placement of the tools, bones, and so on, it can be seen afterwards, since all significant features are noted by the three space coordinates, as well as in relation to the visible stratigraphy. A tool, for in-

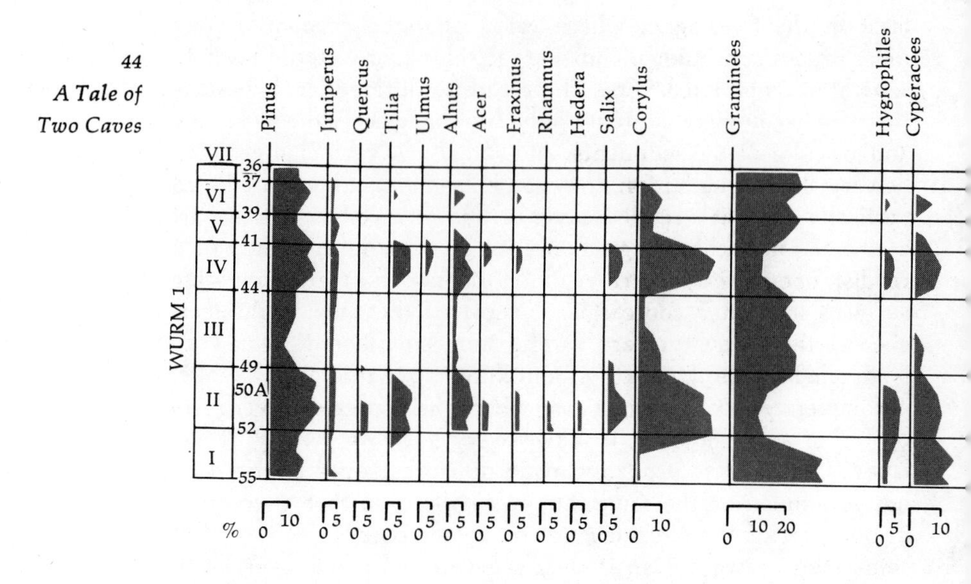

Fig. 13 *Pollen diagram of Würm I at Combe-Grenal.*

The percentages of the different pollen types (species, genus, families) are calculated in respect to the total of pollens (trees + grasses), fern spores being counted apart. These percentages yield a pollen spectrum for each level and the results are then represented graphically. A graph is drawn for each genus or family. The total makes up the pollen diagram of the level studied. One can read, vertically, the evolution of an element of flora during the deposition; or, horizontally, the composition of the flora at a given level. Often, as is the case here, certain elements with very similar ecological characteristics are put together and represented by the same graph. For instance, under "hydrophilous plants" are grouped plants growing in a damp environment (the banks of a river, swamps, etc.), while under "heliophilous plants" are grouped those characteristic of sunny open spaces. "Steppic plants" groups plants growing only on very dry lands or steppes. The ratio of arboreal pollens (A.P.) to nonarboreal pollens (N.A.P.) indicates the percentage of trees.

The composition of the flora at any given point in the diagram indicates the general characteristics of the contemporary climate. Evolution of the different elements (variations in percentages, A.P./N.A.P., presence or absence of certain types) indicates the climatic evolution during deposition of the sediments.

For instance, on the pollen diagram of Würm I in Combe-Grenal, variations in the pollens characterize seven horizontal zones, corresponding to seven successive climatic phases. These can be analyzed as follows:

Zone I, layers 55 to 53: *The arboreal ratio (A.P./N.A.P.) is low (15 to 12%). The Scotch fir* (Pinus sylvestris) *dominates in the trees*

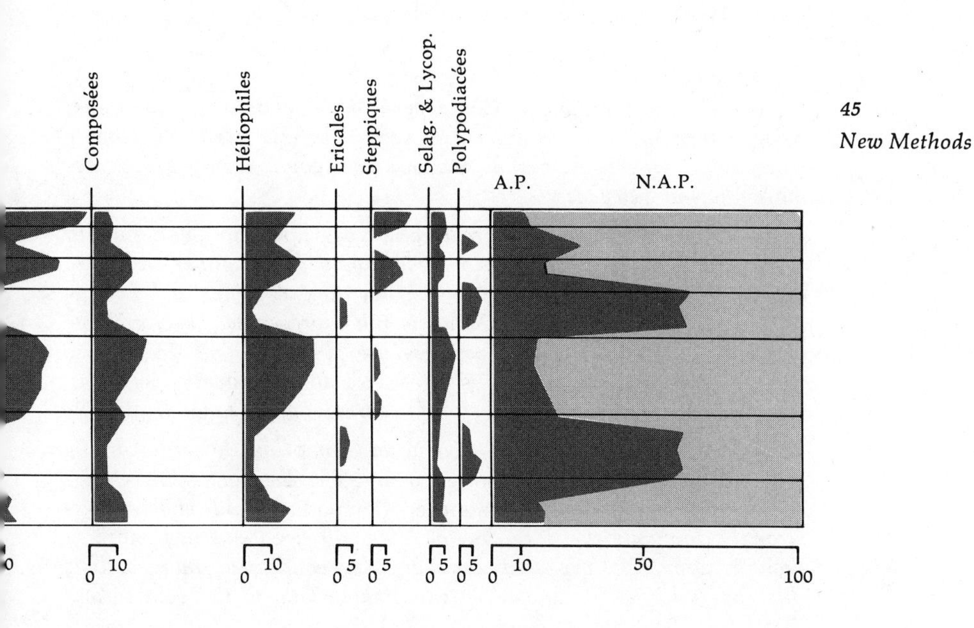

Fig. 13 (cont.)

(some sporadic deciduous trees: alders, willow, hazels). The grasslike plants are dominated by the Graminaceae, *the* Composeae, *and numerous Heliophilous. This flora indicates a cold climate, with some clumps of pines surrounded by wide dry meadows. But the presence of some hydrophilous plants indicates the persistence of certain damp spots in the landscape.*

Zone II, layers 52 to 50A: *The arboreal ratio goes up, very strongly, to 60%. Besides the Scotch fir there are numerous thermophilous deciduous trees (hazel, alder, willow, elm, linden, oak, maple, ash). Among the grasslike plants, Hygrophilous and* Cyperaceae *are very numerous, as well as fern spores of temperate climate* (Polypodiaceae). *All this indicates a strong climatic amelioration, a temperate and damp climate, with light mixed forests, important hazel copses, and numerous swampy zones.*

Zone III, layers 49 to 44: *The arboreal ratio falls to 14%. The deciduous trees are in regression, the most thermophilous disappear. Again,* Composeae *and Heliophilous are in expansion, and some Xerophilous and steppic elements appear. All this indicates a dry and cold climate, drier than zone I which had no steppic element.*

Zone IV, layers 43 to 41: *The arboreal ratio climbs once more to about 60%. As in zone II, the deciduous trees are numerous, and thermophilous elements reappear. Also, hygrophilous elements,* Cyperaceae, *and temperate ferns multiply. Here again we have a temperate and damp phase, with quite a dense tree population, as in zone II.*

Zone V, layers 40 to 39: *The arboreal ratio falls to 13%. Correlatively, all the deciduous trees disappear progressively. The only trees*

Fig. 13 (cont.)
are Scotch firs. Gramineae, Composeae, *and* Heliophilous *dominate; steppic elements are numerous. A very cold and very dry climate corresponds to this flora. The increase in steppic elements indicates more and more dryness.*

Zone VI, layer 38: *The arboreal ratio goes up again, but only to 33%. Scotch fir is in progression, accompanied by juniper and quite abundant hazel; there is a certain percentage of alder and willow. Linden and elm are sporadic. Oak is not represented. Hygrophilous elements,* Cyperaceae *and* Gramineae *are abundant, but* Composeae *remain numerous enough. This flora indicates a climatic variation, less cold and damper, but much less warm than in zones II and IV.*

Zone VII, layers 37 and 36: *A new and important lowering of the arboreal ratio, which does not go over 10%. Deciduous trees have completely disappeared.* Gramineae, Composeae, *and* Heliophilous *strongly dominate the flora. Steppic elements are numerous, more so than in zone V. Very cold and very dry conditions, more steppic than in zones I, III, and V, predominate. This is the coldest and driest episode of the sequence.*

stance, will be entered the following way in the excavation notebook: In square A-10

No. 32 25 145 76 side scraper Top of layer D Horizontal

No. 32 is the order of discovery inside the square A-10; 25 is the distance in centimeters of the middle of the tool from the left side of the square; 145 is the depth of the base of the tool below the datum line; and 76 is the distance from the rear side of the square (the one which is closer to the excavator). Top of layer D refers to its position inside the layer, and horizontal means that the tool was not tilted.

At the same time, the tool is drawn to scale on a plan of square meter A-10 at the depth of, let us say, 142 to 146. The juxtaposition of such plans permits a later reconstruction of the living floor, if such a floor were discernible.

By using this kind of notebook, any kind of section can in fact be drawn for any position parallel to the sides of the squares, or even, with some complications, diagonal to the square (Fig. 14). These sections can be compared with the geological sections taken during excavations, and one can thus follow all the variations in thickness of the layers. Horizontal maps can also be made, of course.

One should be careful when using only the color of a layer to separate it. In the first year of excavations at Combe-Grenal, we

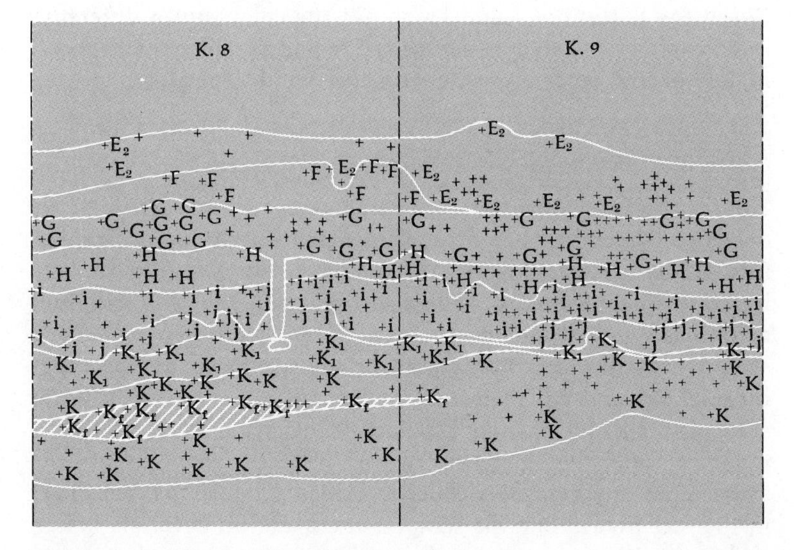

Fig. 14

Construction of a section by projection of coordinates. This is a sagittal-position diagram, that is, a diagram obtained by projection of two of the coordinates, the depth and the sagittal coordinates. It is a 0–25 cm diagram, which means that only the tools whose frontal coordinates were between 0 and 25 cm in value were taken into account. Each cross represents either a flint or a bone. In square K-8 one can see the section of the posthole. The shaded area corresponds to a sublayer rich in ashes inside layer K. (Kf = K foyer, "foyer" *being the French word for hearth.)*

used the color to separate layer J (now 20), which was blackish, from layer I (now 17). (There are two intermediate layers deeper in the cave, both gray.) The analysis of the assemblages then showed us that there was something wrong; J is Denticulate Mousterian, I Quina Mousterian. But in this first excavation, some Quina-type scrapers occurred in the assemblage of layer J, which was surprising since it was the only case in which such scrapers had been found in Denticulate Mousterian. So, the following year we were more careful about the limit between I and J, and discovered first the existence of numerous microgullies at the top of J (there had clearly been erosion by small streams), and then that the bottom of I was colored in black by redeposited ashes from J. Every time we discovered a Quina scraper that did seem to belong to J, it was either in one of the microgullies, or in the black part of I. The true difference between the bottom of I and J was not the color, but a difference in texture: J consisted of coarse sands and small congelifracts, I mainly of fine sands and silts, probably wind-deposited.

Methods of Studying Assemblages

For the study of Mousterian assemblages, as early as 1949 we developed a method based on elementary statistics, which takes into account the totality of the artifacts rather than a few selected as "characteristic." We also used technological attributes as well as typological ones.

A Mousterian assemblage can be characterized by the types represented and by their relative proportions (the typological side), and also by the way the implements were made (the technological side). For instance, two assemblages may contain the same types of scrapers in about the same proportions, but differ in the fact that in one the scrapers are made on Levallois flakes, and in the other on ordinary flakes. The type of retouch can also play a part in differentiation.

For the typological study, we use a list of 63 different types of tools, some divided into subtypes, plus a list of 21 types of handaxes (when present). Cores, ordinary flakes and blades, special flakes obtained in the fabrication of handaxes, and chips are all counted apart, since they are not "tools" in a strict sense, even if they could have been utilized as such. An example of this typological list can be found in the appendix.

For the technological study, we use a special card (Table 1), divided into two main categories: Levallois and non-Levallois. Each of these two-category frames is horizontally divided into three: flakes, points, and blades ("points" should be taken here in the technological sense, that is, triangular flakes). Vertically, each of the frames is divided into six columns, following the type of butt on the flake: plain, faceted, convex-faceted, convex-dihedral (only two facets forming an angle), taken away by retouch, and broken or not recognizable. Cortex-covered butts are classified as plain.

The analysis process is very simple in theory. Let us suppose we have beside us all the artifacts found in a layer. The initial one, A, that we examine is a scraper. We determine first its technical characteristics. The flake on which it has been made is Levallois, so it will be counted in the upper frame. Its length is more than twice its width, and was probably so before the retouch, so it will go in the bottom row of the upper frame. The butt is faceted and convex at the same time, so it will go in the third column (see Table 1). Typologically, it is a side scraper,

Table 1

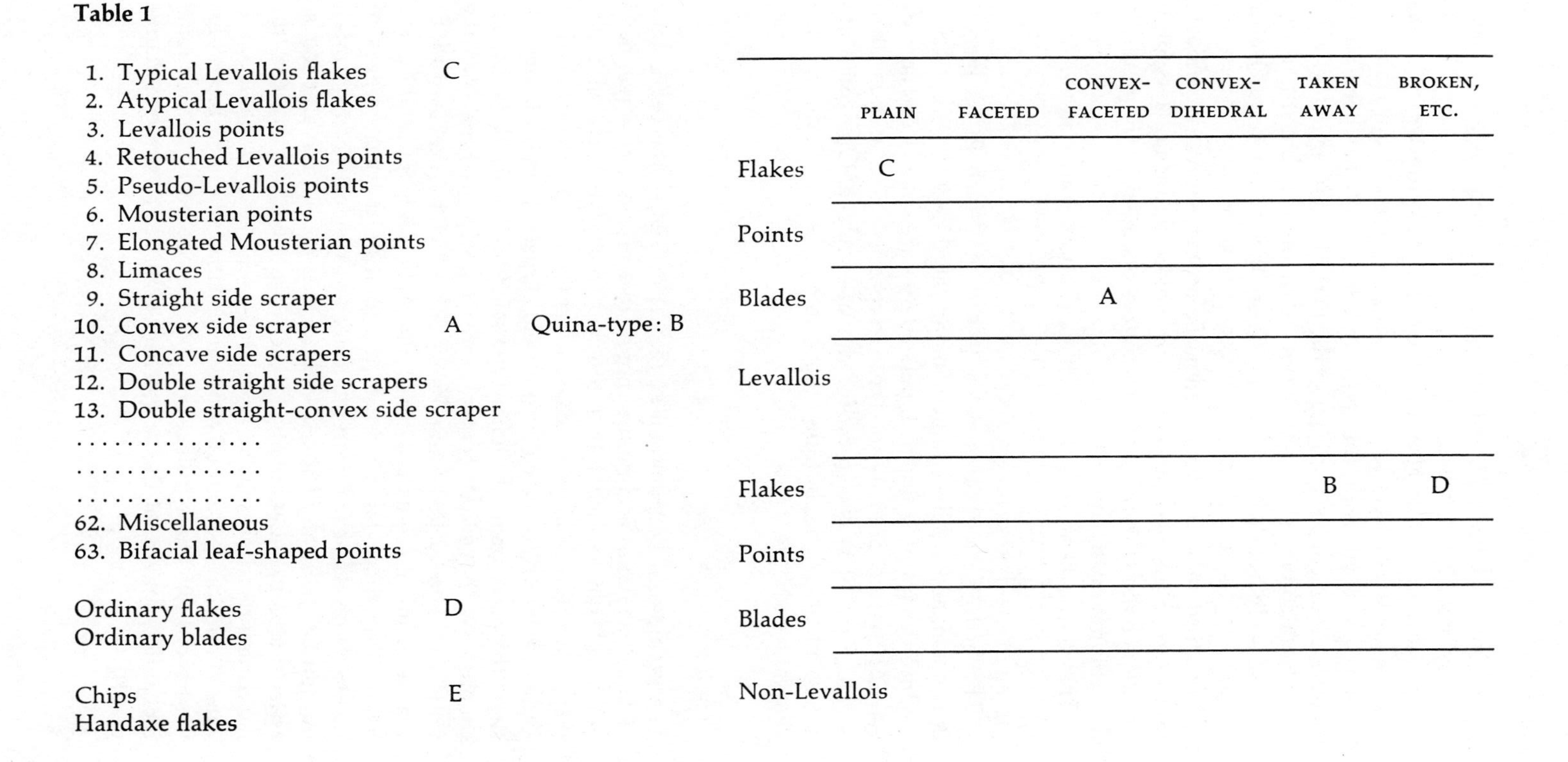

1. Typical Levallois flakes	C	
2. Atypical Levallois flakes		
3. Levallois points		
4. Retouched Levallois points		
5. Pseudo-Levallois points		
6. Mousterian points		
7. Elongated Mousterian points		
8. Limaces		
9. Straight side scraper		
10. Convex side scraper	A	Quina-type: B
11. Concave side scrapers		
12. Double straight side scrapers		
13. Double straight-convex side scraper		
. . . .		
. . . .		
. . . .		
62. Miscellaneous		
63. Bifacial leaf-shaped points		
Ordinary flakes	D	
Ordinary blades		
Chips	E	
Handaxe flakes		

		PLAIN	FACETED	CONVEX-FACETED	CONVEX-DIHEDRAL	TAKEN AWAY	BROKEN, ETC.
Levallois	Flakes	C					
	Points						
	Blades			A			
Non-Levallois	Flakes					B	D
	Points						
	Blades						

with a convex edge, so it will be counted among the convex side scrapers, no. 10 of the list.

The next tool, B, is also a scraper. The flake on which it has been made is not Levallois, so it will go in the bottom frame. It is short, so it will go in the top row, and its butt has been taken away by retouch, so it will go in the fifth column. Typologically, it is also a convex side scraper (no. 10 of the list) but it is thick, with a scaly retouch, so it will go in the subdivision Q (for Quina).

The third tool, C, is a Levallois flake that shows no retouch, except some traces of utilization. It will go under no. 1 (typical Levallois flake, unretouched). Its butt is plain, so it will go in the first column of the upper frame.

The fourth, D, is an ordinary flake (bottom frame, upper row) with a broken butt (last column). On the typological side, it is placed in the flakes category.

The fifth, E, is a small flake, less than 1 inch in its largest dimension, and will be counted among the chips.

Once the count is finished, percentages are calculated for each type of tool on the total of the 63-types list, and a cumulative graph (Fig. 15) is drawn up for this assemblage. One could use any kind of graph, but the cumulative one best indicates the general trend.

Typological Indexes

The typological Levallois index (TyLI) expresses the percentage of Levallois flakes and points (nos. 1 to 3 of the list) which have not been transformed into retouched tools. The scraper index (SI) shows the total percentage of scrapers (nos. 9 to 29). The Quina index (QI) is of a different type, being designed to show the percentage of Quina-type scrapers *among the scrapers.* (Suppose we have 100 scrapers, and 18 are Quina-type, then the QI is 18.) The backed-knife index, or unifacial Acheulean index (backed knives are as typical of the Acheulean tradition as the handaxes), abbreviated UAI, is the total of the percentages of no. 36 (typical backed knife) and no. 37 (atypical backed knife), but not 38 (naturally backed knife, which is found in any Mousterian industry). Eventually, if there are handaxes present, the handaxe index (HI) will be calculated as the percentage of handaxes in relation to the total of the tools (that is, the 63 types plus the handaxes). These indexes can be represented by proportional rectangles, drawn on the same sheet of paper as the cumulative graph.

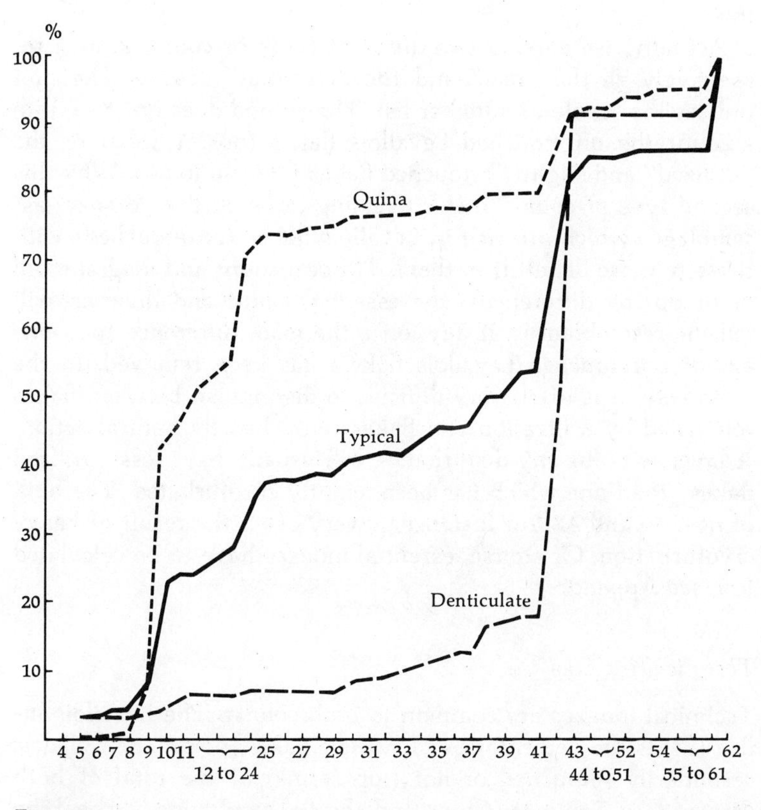

Fig. 15

Examples of cumulative diagrams—three different types of cumulative diagrams are represented: Quina Mousterian, Typical Mousterian (subtype relatively poor in scrapers), and Denticulate Mousterian.

It is also convenient to calculate other types of indexes, known as "characteristic groups." Group I, or the Levallois group, comprises nos. 1 to 4. Group II, the Mousterian group, comprises nos. 6 to 29. The third group, III, comprises the types that will show great development in the Upper Palaeolithic, and so is called the "Upper Palaeolithic group." It contains nos. 30 and 31, typical and atypical end scrapers; 32 and 33, typical and atypical burins; 34 and 35, typical and atypical borers; 36 and 37, typical and atypical backed knives; and 40, truncated flakes and blades. Group IV is made up of one type only, the denticulates (no. 43), but in fact this "type" covers several subtypes of tools.

Actually, we employ two different types of count, known respectively as the "real" and the "essential" counts. The first utilizes the whole 63-number list. The second does not take into account the unretouched Levallois flakes (nos. 1 to 3) or the "utilized" and slightly retouched flakes (nos. 46 to 50). Why this second type of count? It is interesting to be able to compare assemblages which are rich in Levallois flakes (unretouched) with those that are deficient in them. The real count and diagram will point up the differences; the essential count and diagram will tell the resemblances, if any, once the main difference (percentage of unretouched Levallois flakes) has been removed. In the same way, it is often very difficult to distinguish between flakes retouched by utilization and flakes retouched by natural action. A layer without any disturbance will usually have less "utilized flakes" than one which has been slightly cryoturbated. The bulk of nos. 46 and 47, for instance, is very often the result of heavy cryoturbation. Of course, essential indexes have to be calculated too (see Appendix).

Technical Indexes

Technical indexes are common to both counts. The Levallois index (LI) is the percentage of Levallois flakes, points, and blades, secondarily retouched or not (top frame) of the total of both frames (see Table 1). Chips and the flakes obtained in making handaxes are excluded here.

The faceting index (FI) is the percentage of faceted butts (total of columns 2, 3, and 4 in the two frames on the total of recognizable butts, columns 5 and 6 excluded). The restricted faceting index (FI^r) is the percentage of butts with small facets (in which case column 4 is also excluded).

The laminary index is the percentage of blades, Levallois or not, in respect to the total of the two frames.

Diagrams, and typological and technological indexes, enable us to gather the main characteristics of an assemblage on a single sheet of paper. Comparison with another assemblage is then easily done by transparency.

This type of analysis requires a good knowledge of typology and techniques, but it is interesting to note that students, provided they follow the typological definitions of the tools, very quickly get diagrams which are close to those obtained by experts.

The use of the analysis methods outlined above on the Mousterian assemblages of southwest France led in fact to the recognition of several Mousterian industries, which are now known as the Mousterian complex.

There are three main types of cumulative graphs (Fig. 15). The first characterizes assemblages rich in side scrapers (more than 50 percent) and low in denticulates: Quina-type, Ferrassie-type, part of the Typical Mousterian. The second characterizes assemblages with a moderate percentage of side scrapers and a rather higher percentage of denticulates than in the preceding type: Mousterian of Acheulean tradition, type A, and part of the Typical Mousterian. The third has a low percentage of side scrapers (from about 4 to 20 percent) and a very high percentage of denticulates (up to 60 percent) and notches: Mousterian of Acheulean tradition, type B, and Denticulate Mousterian.

Combining the cumulative graphs with the indexes, we arrived at the following classification of Mousterian industries in the southwest.

1. *Charentian Group* (so named because it is strongly dominant in the district of Charente). This is divided into two subtypes:

(a) Quina subtype: very high scraper index (up to 80) usually, but sometimes falling, toward the end, to 50. A high proportion of transverse scrapers (nos. 22 to 24 of the list). A high Quina index (14 to 30), with an absence or rarity of true handaxes (which when found are of special types) or backed knives. Very low value of group IV, except toward the end, where it is simply low. Very low Levallois index (less than 10, often around 2 or even 1).

(b) Ferrassie subtype: very high scraper index (comparable to Quina) but a rather low percentage of transverse scrapers. A medium Quina index (6 to 14), with an absence or rarity of handaxes (which when found are of special types) and backed knives. High Levallois index (14 to 30).

Some assemblages range between these two subtypes, which seem to be just variants, facies, of the same general type.

2. *Typical Mousterian Group.* The scraper index is variable and seems to divide this group into two subtypes. A rather low percentage of transverse scrapers, a low Quina index (often 0, maximum 3), and an absence or rarity of true handaxes and backed knives. Levallois index very variable.

3. *Mousterian of Acheulean Tradition Group.* This is again divided into two subtypes:

(a) Subtype A: variable scraper index, but never very high or very low (25 to 45). Quina index very low or null. Denticulate often fairly numerous. A variable handaxe index (seldom lower than 8, and sometimes up to 40). A variable percentage of backed knives, but the UAI is seldom over 4. Group III (Upper Palaeolithic) often fairly well developed. Levallois index very variable.

(b) Subtype B: low scraper index. Quina index null. A low handaxe index, and the handaxes often look "degenerate." A high percentage of denticulates. A high percentage of backed knives, often on blades. Group III well developed. Variable Levallois index. This subtype is chronologically younger than subtype A and derives from it. Intermediates are known.

4. *Denticulate Mousterian Group.* This has a low to very low scraper index, often with "degenerate" scrapers. The Quina index is either very low or more usually null. A high to very high percentage of denticulates and notches. An absence of true handaxes and absence or extreme rarity of backed knives. Variable Levallois index.

All these types and subtypes exist either at Combe-Grenal or Pech de l'Azé, and many exist at both sites. Later, we shall discuss their significance.

3. *Back to Pech de l'Azé*

At this point it is helpful to return to the 1967–1969 excavations and their results at Pech de l'Azé II. All the methods outlined above have given us a great deal of new information. But these excavations did not bring about any major change in the stratigraphy that had been established in 1952, with one notable exception—the discovery of Riss III layers inside the cave.

As you may remember, we had been rather surprised to find only what were apparently the two older members of the Riss sequence. In some places, small remnants of reddish eboulis did exist between the upper Rissian and early Würm levels, but it was not much to justify assuming a third Rissian subdivision. When the excavation got into the cave itself, however, we found —between the soil at the top of Riss II and the yellow, unweathered eboulis of Würm I—a thick (0.80 m) layer, reddish and weathered, which could well represent deposits of the Riss III period. Unfortunately, this layer is almost sterile, containing only some bones and a minimum of flints; but the pollens fit in very nicely with the Riss III at Combe-Grenal.

THE PALAEOLITHIC SEQUENCE AT PECH DE L'AZÉ II

Under the Rissian layers, the earlier deposits seem absolutely sterile. The pollen analysis is not yet completed, but it appears that the upper part at least belongs to the great interglacial, the Mindel-Riss. These sands and gravels, well-bedded, suggest a quite rapid stream and probably date back to the time when the cave was uninhabited and closed. If so, we could date the erosion by the valley, which opened the cave to human occupation, to the end of this Mindel-Riss interglacial, since the cave was occupied as early as Riss I. Or perhaps the opening came a little later, which would explain why the very early Riss stage (pre-Riss) is not found at this site.

The Rissian Layers

Riss I is divided into two parts: the lower part, layer 9; and the upper part, "layer" 8, which is not a true layer, but just a weathering, a soil. But as it is regular enough on most of the excavated part, it can serve as an artificial stratigraphical subdivision.

Layer 9 is composed of rounded eboulis, from 1 to 5 inches in diameter, mixed with sands and disposed in irregular pockets. This layer seems to have been heavily disturbed by cryoturbation and/or running water. However, in some protected places traces of fire can be seen. Judging by the sedimentology and the pollens, it was deposited under very cold and dry conditions. The only trees existing were a few pines, most of the vegetation being a grassy steppe. The fauna, taken alone, would argue "warm" conditions. It consists of an abundance of red deer, the roe deer, the great Irish deer (*Megaceros*), numerous bovids, at least one elephant (probably *Elephas antiquus* from its size), numerous horses, Merck's rhinoceros, wolves, rabbits, and so on. The presence of red and roe deer seems to indicate that some trees at least existed in sheltered parts of the country.

Layer 9 yielded 330 artifacts, 113 of which are retouched tools. The flakes and tools were often battered by cryoturbation and difficult to characterize, except for the handaxes. The Levallois index was very low (3.5) and the butts of the flakes seldom faceted (FI = 35.5). Very few "blades," and these are rather elongated flakes rather than true blades. Side scrapers were not very numerous (SI = 20) and usually rather poor. End scrapers, both typical (6 percent) and atypical (8 percent), were strongly represented, but it may be that some of them are the result of natural actions. It may be the case also with some of the borers (5 percent). Burins and backed knives account for 4 percent each. Notches and denticulates are numerous, but here too some may be the result of natural action. One flint chopper, and three chopping tools made of quartz were found. There were 10 handaxes (HI = 9.1), rather crude, which found out of context would probably be attributed to Abbevillian rather than Middle Acheulean. The raw material (quartz, basalt, and other eruptive rocks) is partially to blame, but the flint ones are not much better (Fig. 16). Note: a flake cleaver was found close to a quartz handaxe.

The flakes often have plain, inclined butts, with big percussion bulbs and cones. They have been flaked with a stone hammer, and very often the retouch was also done with a hard hammer.

Layer 8 is a reddish brown soil, which represents the weathering of the top of Riss I deposits during the following interstadial. The sediment was deposited under wetter and less cold conditions, and it was probably at the onset of these conditions that layer 9 was cryoturbated. The vegetation is richer. Most of the trees are pines still, but there are some birch and hazel trees too. Red deer is abundant, also bovids and horses. In addition, there are roe deer, *Megaceros*, Merck's rhinoceros, a primitive type of cave bear, wolves, badgers, rabbits, and so on.

The artifacts are a little more numerous, 560 in all, 228 of which can be classified as tools. In general, there is little difference from layer 9. There are four choppers, seven chopping tools, and ten handaxes. Among the handaxes, there is one flake cleaver made of quartz. The cores are shapeless, and the flaking technique is very much as in the preceding layer.

The top of Riss II is weathered to a variable depth. Outside the cave, the weathering is deep, and only some 4 inches of unweathered material remain at the bottom. Inside the cave, this weathered zone tapers off and disappears. Of course, this pedological horizon has nothing to do with either geological or archaeological layers, and it would be a big mistake to speak of the "red layer" and the "yellow layer." It is interesting to note that this soil penetrated less deeply into the cave than the soil of the first Rissian interstadial, now much thinner. Possibly a good part of this first soil was eroded away before the deposition of the Riss II layers.

The lower part of Riss II has seen a more temperate, and much damper, climate than Riss I. This may have been the last part of the interstadial. It was a park landscape (about 25 percent tree pollen), with clusters of pines, birches, hazels, alders, and willows, and even some warmth-loving species like elms, linden, beech, and oak, the last very rare. Among the grasses were plenty of damp-climate types.

Toward the top, the climate becomes progressively drier and colder. The landscape is one of dry prairies and steppes, with some pines, sporadic hazels and birches, sometimes alders and willows. Grasses thriving under dry conditions replace the damp-climate ones.

The lower archaeological layer, c, has yielded only 321 artifacts, 116 of which are retouched tools. The industry seems still battered by cryoturbation. The Levallois index is low (8.5), and the butts are seldom faceted (FI = 34.5). There are, however, three typical Levallois flakes. The side scrapers are in about the same percentage as before (SI = 22.4) and belong to sundry types. Here, too, the Upper Palaeolithic types of tools are numerous: end scrapers (9.3 percent); burins (1 percent); borers (2.8

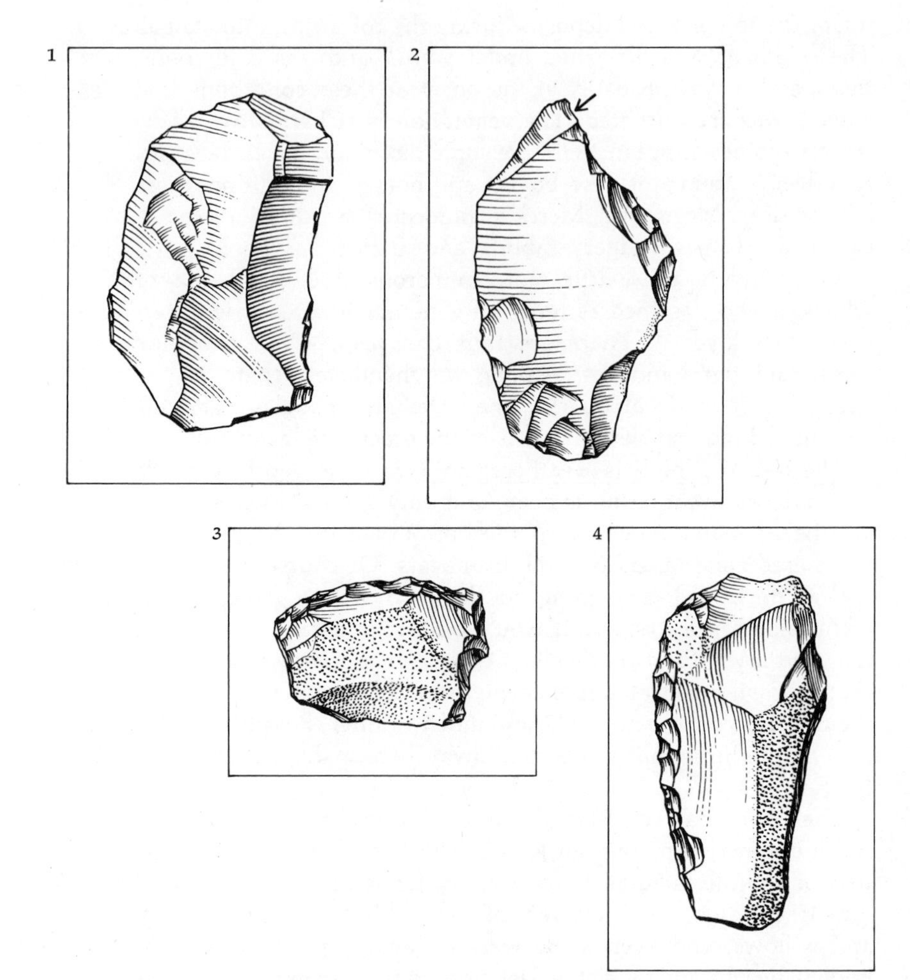

Fig. 16

Acheulean at Pech de l'Azé II.

1, Levallois flake (Riss II, toward top). 2, Burin. 3, Transversal scraper. 4, Convex side scraper (2 to 4, Riss II, bottom part). 5, Basalt handaxe (Riss I).

percent); typical (2.8 percent) and atypical (3.7 percent) backed knives; truncated flakes (4.7 percent); in all about 24 percent, more than the side scrapers. Notches, denticulates, and so on, complete the flake-tools kit. Five choppers, four chopping tools, and five handaxes (HI = 4.5)—one a short lanceolate, one a "backed" handaxe, one a truncated amygdaloid, and the remaining two crude, more or less Abbevillian-like (Fig. 16).

5

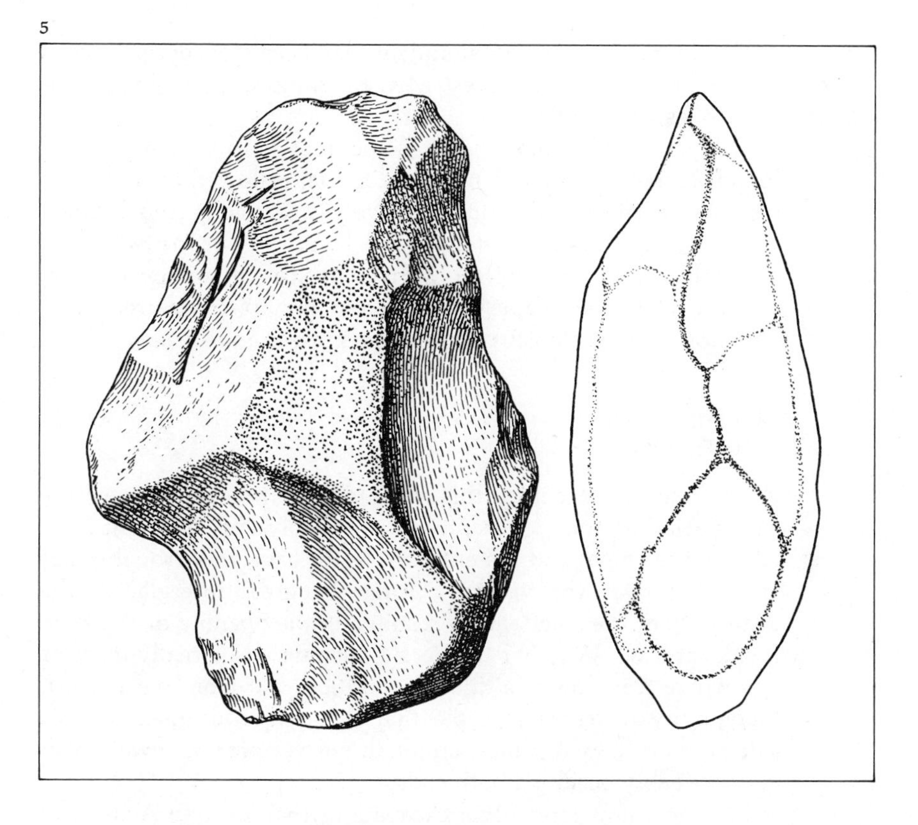

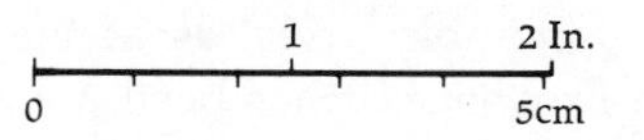

The techniques of flaking are close to that used during Riss I.

In the fauna, both red and roe deer are well represented, as well as bovids and horses. Cave bear is frequent. Also Merck's rhinoceros, the wild boar, wolves, hyenas, panthers, and so on are represented.

Layer b has yielded only 167 artifacts, of which 62 are tools. This is a pity, for there seems to be an important change in the industry. The Levallois index jumps to a value of 40.4, and many flake butts are now faceted (FI = 64). Some flakes seem still to be battered by cryoturbation; unhappily, the small number of tools somewhat lowers the significance of the typological changes. The scrapers are more numerous than the Upper Palaeolithic tools. Notches and denticulates are well represented. There is also one enormous basalt chopper, two chopping tools, and only one handaxe (1.6 percent), but this may represent a statistical fluctuation (Fig. 16).

The fauna consists of abundant red deer, roe deer, bovids, wild boar, cave bear, horses, Merck's rhinoceros, wolves, bobcats, foxes, and rabbits.

In layer a, only some flakes (sometimes Levallois) have been found. Horses are abundant in the fauna, followed by red deer, roe deer, a species of wild goat, wolves, badgers, and rabbits.

Riss III is practically sterile. The layer was deposited under dry, cold conditions, the top part being drier than the bottom. The prevalent landscape was a steppe with rare pinetrees, and numerous steppe elements among the grasses and shrubs.

PALETHNOLOGICAL OBSERVATIONS ON THE ACHEULEAN AT PECH DE L'AZÉ

When man first occupied this cave, toward the beginning of the Riss glaciation, the topography was probably very different from what we see now, and the entrance to the cave was probably not very high over the valley. During the great interglacial, the deepening of the small valley resulted in the opening of the Pech II side, at least. Whether the Pech I opening was already more or less where it is now is a difficult question, but from the modern topography it seems possible that the cave continued on this side also for some distance, although not very far. Anyway, man very probably occupied both ends.

We know now from other excavations that, even in Acheulean times, living space under the vaults of the caves was organized. In especially favorable cases, as at the Lazaret (25) and Lunel Viel caves in Provence, it has been possible to find the traces of huts, or tents, and even to tell where the animal skins which were used as beds lay, from the concentrations of small wrist bones, and so on. In Pech de l'Azé, alas, the effects of running water, even if a mere trickle most of the time, have destroyed this possibility. But some other features are present. For example, numerous fireplaces have been found in Pech de l'Azé II, at different levels in the Rissian layers (Fig. 17). Some are situated well inside the present cave, and must have been even deeper in Acheulean times since part of the roof collapsed later. These hearths are of three different types. First, elementary hearths, which are just traces of fire, reddened sands, and black ashes, without any special preparation. They are usually small, but sometimes more than 1 meter wide, and roughly circular in shape. They are thin and indicate fires which did not burn for very long. Sometimes some stones are disposed around the blackened areas.

Second are paved hearths. These have been lighted over a

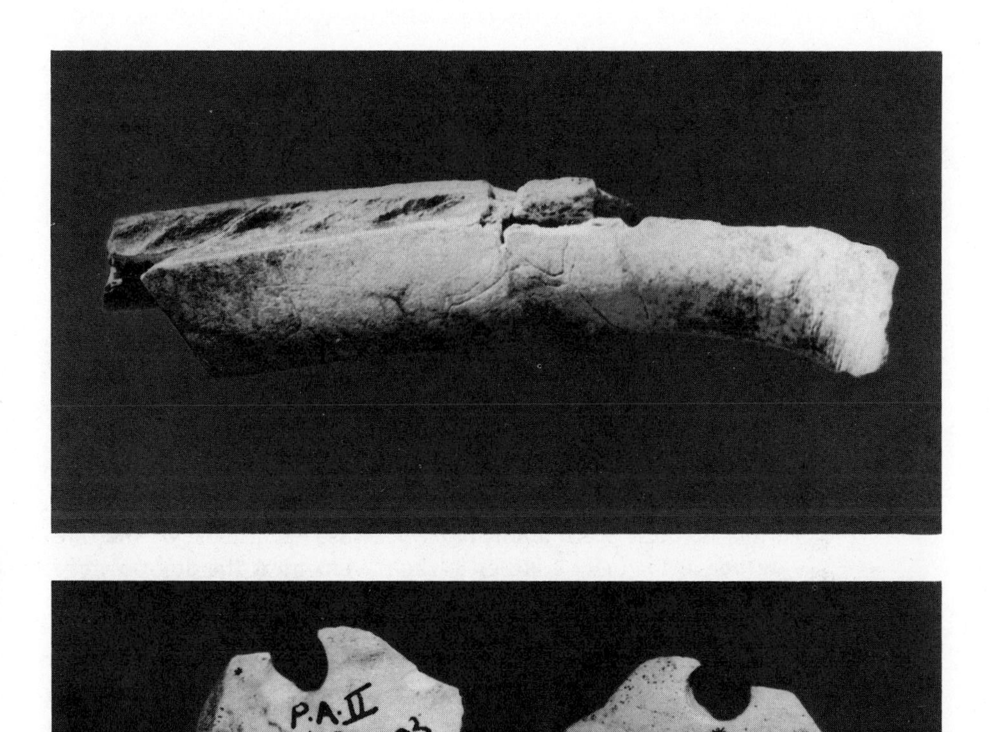

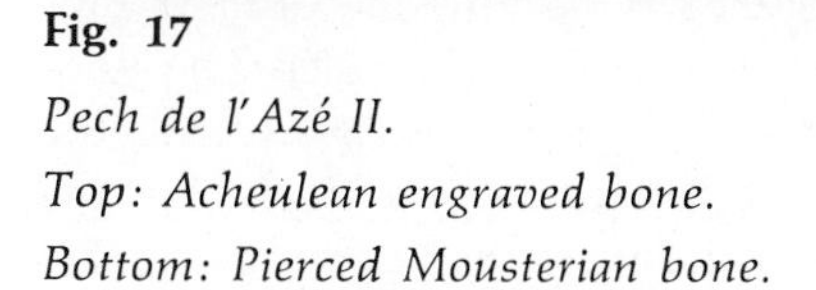

Fig. 17

Pech de l'Azé II.

Top: Acheulean engraved bone.

Bottom: Pierced Mousterian bone.

pavement of flat stones which have been reddened, and can reach over 1 square meter. The intense reddening of the stones seems to indicate they have been used for a long time. Perhaps they were cooking places, the stones being heated by a hot fire, then brushed clean and the meat laid on them.

Third, dug-out hearths. These are the kind with a "channel" described by Pavel Boriskovski in the Russian Upper Palaeolithic site of Kostienki XIX, on the Don River, and found since by Eduardo Ripoll and myself in the Upper Solutrean of Cueva de Ambrosio in Andalusia (southern Spain). I also found them in the Aurignacian I of the Roc de Combe cave and in the open-air site of Corbiac (both in southwest France)—the latter in a very well-evolved Perigordian. But here at Pech de l'Azé II, we find them dating from many millennia earlier! They are small (about 6 to 10 inches in diameter) and were dug into the sand, which is more or less reddened all around (Fig. 18). They were filled with black ashes. In Pech de l'Azé II, they appear with level b in Riss II, at the same time that there is a change in the tool kit. These two simultaneous changes could represent the coming of a slightly different Acheulean tradition, but the archaeological data are unfortunately insufficient to confirm this.

The disposition of the hearths is interesting. As I said, some were deep into the cave, but others were closer to the entrance. Most of these were amorphous hearths, while most of the ones deeper inside belong either to the pavement or the dug-out type.

At the bottom of the Riss I deposits on the right side (looking inside) lay a pile of bones, and just beside it a huge red deer antler. It is difficult to tell if this was intentional or just the result of sweeping out the bones from the passage and the living place itself.

Toward the top of "layer" 8 a very interesting discovery was made, although it was not recognized as such immediately (9). We found an engraved bone. (The oldest known before this discovery belonged to the beginning of the Upper Palaeolithic, at least 100,000 years later.) It is the proximal part of an ox rib and carries, on its flat side, near the lateral groove of the rib, a series of lines and incisions which are clearly intentional, not the random lines left by a flint cutting off the meat (Fig. 17). The lines are often double but not parallel, which seems to exclude the possibility of a double-pointed tool. It seems to be by far the oldest engraving known, even if the work has no meaning for us. Was it done by an idle Acheulean hunter to amuse himself? Was it the first tentative effort to represent something? Or was it a symbolic drawing? Probably we shall never know.

Anyway, the occupation of cave II seems to have been discontinuous. It is less well exposed than cave I, and by the look of the deposits and the numerous channels dug by running water into the layers, even in the Riss II drier times, it seems to have been quite damp. It was probably occupied only in summer, the main occupation being on the other side. Tools are scarce at any level, even if the animal remains are numerous and diverse. It seems, too, from the relative scarcity of unworked flakes, that most of the tools must have been flaked elsewhere. However, the

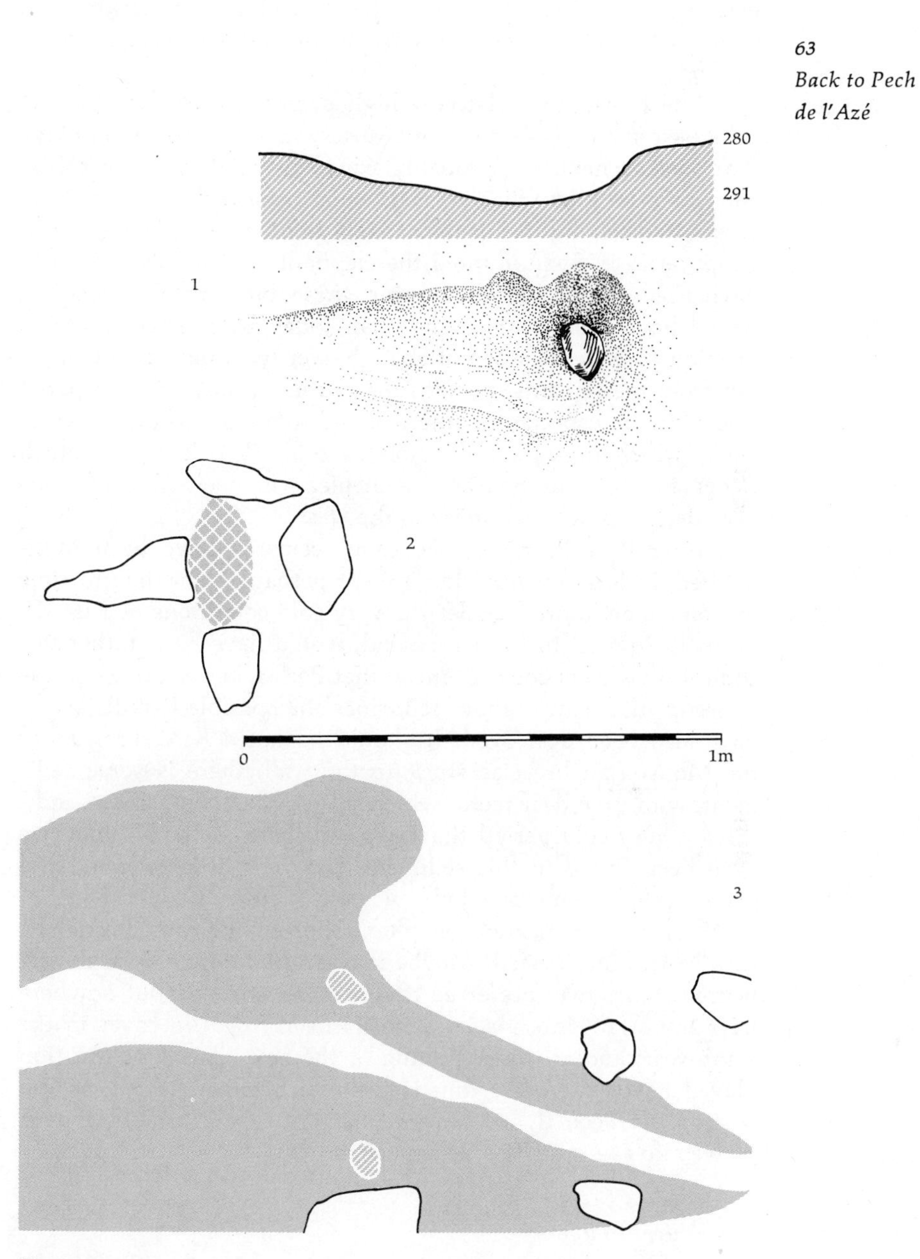

Fig. 18

Hearths at Pech de l'Azé II, Acheulean layers.

1, Section and plan of a "tailed" hearth. 2, Small hearth with stones around. 3, Traces of fire (black ashes, reddened sand), cut by a small rivulet indicating damp conditions inside the cave.

existence of typical handaxe flakes, small retouch flakes, and rare cores indicates that at least some of the tools were made on the spot.

In the fauna, the relatively high percentage of Merck's rhinoceros remains calls for some observations. It seems unlikely that these Acheuleans, probably armed only with wooden spears (there are very few flint points), could have tackled the rhinoceroses in direct fight. But they surely hunted them, as their contemporaries in Spain hunted the elephant, and so were likely to have used traps. Did they bring these big animals home? It would have been made easier than today by the fact that the opening of the cave was not as high over the valley as it is now, but from the remains it seems they brought only selected parts. The teeth are especially numerous and sometimes show traces of wear, which indicate utilization as tools. We know of much richer sites of this Southern Acheulean in open-air conditions near Bergerac, some 60 miles to the west.

During Riss III times, the cave seems to have been uninhabited, at least on this side, except perhaps from time to time by some lone hunter. Under the very cold conditions of Riss III, Pech II, exposed to the west wind, would have been rather uncomfortable. This does not mean that Pech I, better exposed, was unoccupied; but we cannot tell, since the possible Riss III layers have also been destroyed. It should be noted here that, under the Mousterian breccias stuck to the wall, there is some sediment, with cave bear teeth, which from pollen analysis has indicated a very cold period that may well be Riss III. No tool has even been found in this sediment, but very little remains, and even less has been excavated.

We have no indication of inhabitation of the cave during the Riss-Würm interglacial. Maybe it was too damp, maybe during warmer times man preferred to live under the sky. But nowhere have important interglacial deposits been found in caves in the southwest; and at the beginning of the next glacial, solifluction played havoc with any thin layers which might have been deposited. At Pech II, we can see that the first Würm I deposits plowed up even the Riss III deposits.

The Würm I Layers

Würm I begins with layer 5 in Pech II, a huge fall of debris from the roof the cave, perhaps weakened and corroded by chemical alteration during the interglacial. Among these fragments are found large chunks of stalactites, a good proof of damp conditions during the interglacial. Some tools, scattered here and there, testify to the passage of man, but there was no real installation in the mouth of a cave, which was probably very dangerous, the frequence of rock fall being high. This eboulis was formed under a very cold and dry climate. There were very few trees, only pines, in a steppic environment. At the top, the climate became damper and a little less cold, the rate of rock fall diminished, and man occupied the cave again (layer 4e). We know very little of this occupation, for during the beginning of the next period heavy cryoturbation crushed the flints and bones. Only a little part, deep under the shelter, escaped this battering. The tools seem to indicate a typical Mousterian of Levallois flaking technique (Fig. 19). At that time there were more trees, with some birches, hazels, and alders.

The cryoturbated part (called 4d, since initially we thought it was a different layer) contains a lot of tools, but it is not possible to say much about them, since they were heavily battered and bruised by the formation of the polygonal soil. The edges of the flakes have been powerfully crushed and most of the time the true retouches have been obliterated, and pseudo-retouches cover them (Fig. 19). The fauna of this cryoturbated level consists mostly of rabbit and red deer, with some bovids, chamois, rhinoceros (of an indeterminable species), horse, and one bone of reindeer, which was thus present, albeit rare, at the early beginning of the Würm period.

Over this layer of eboulis comes layer 4 proper, consisting of reddish sands and clays, with few congelifracts, and deposited by sheet flow during a damp period. Even if the climate was rather mild all through the layer, there are interesting variations.

Layer 4C2 belonged to Typical Mousterian and was deposited under a damp climate, with a percentage of trees rising progressively to 43; most of them are still pines, with willows, birches, some alders, and hazels. The fauna is represented mainly by red deer (73 percent of the remains), horse (10 percent), and bovids (8 percent), with some roe deer, chamois, rabbits, wolves, and foxes.

Archaeologically speaking, this layer was not very rich; nonetheless, it yielded a total of 2,638 artifacts. The Levallois index is low (9.5), the faceting indexes relatively low also (FI = 50.8, FI^r = 33.5). Most flakes are short, and the laminary index is low, (only 6.0). In real count, the typological Levallois index is low (10.2), and the ordinary flakes and blades which show only traces of utilization or accommodation make up 19.1 of the total.

In essential count (where the Levallois flakes that have not been transformed by retouch and utilized flakes are not included), the scraper index is neither high nor low (SI = 37.8).

1

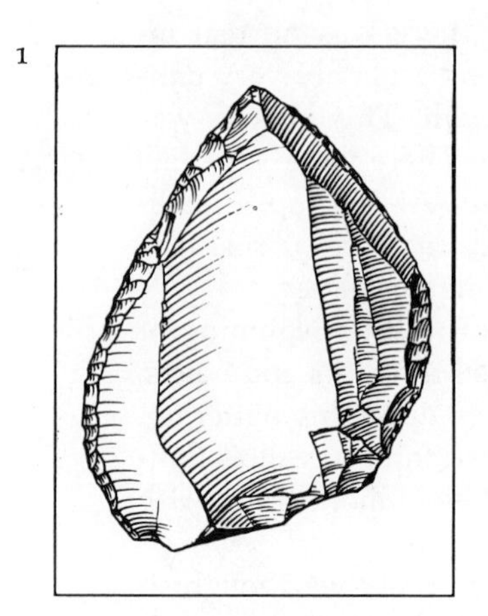

3

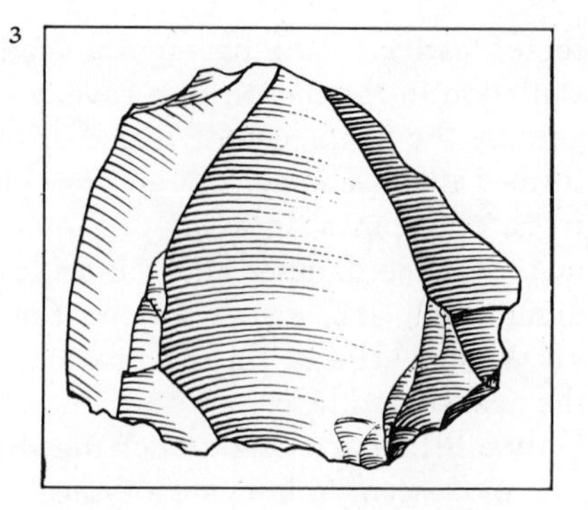

5

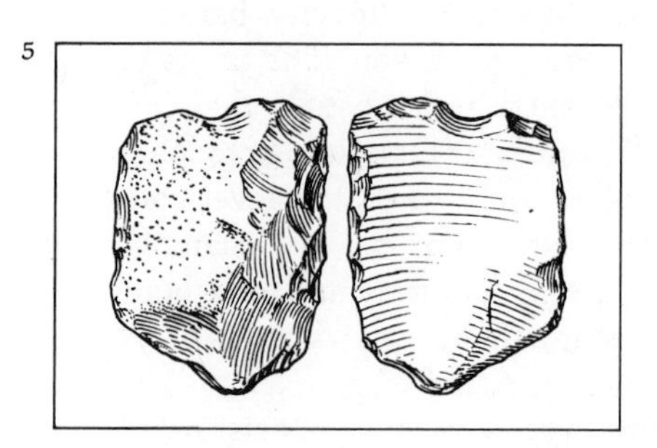

2

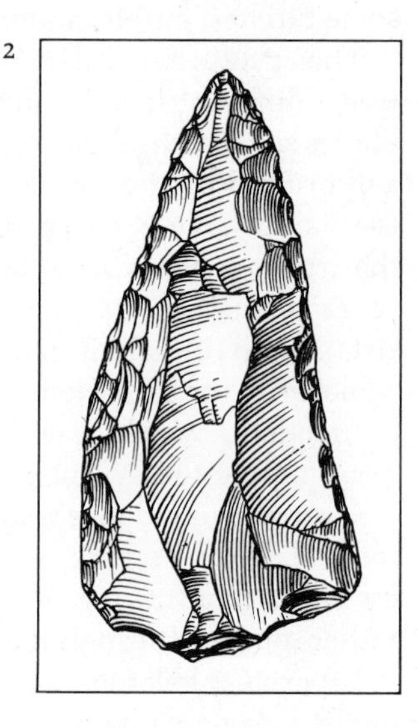

4

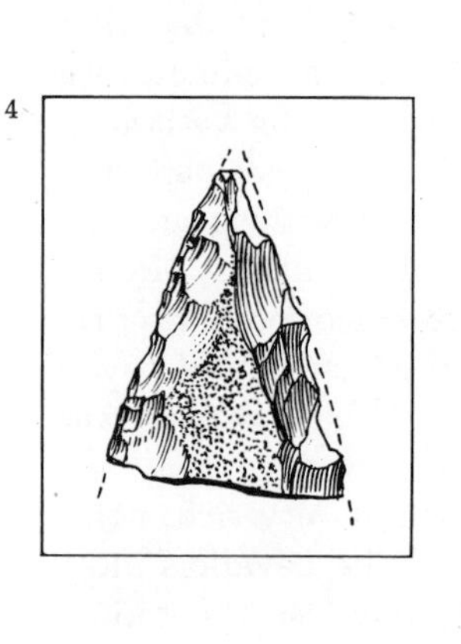

6

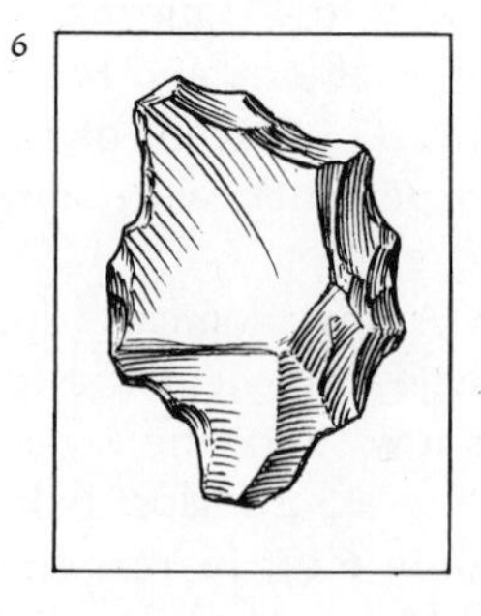

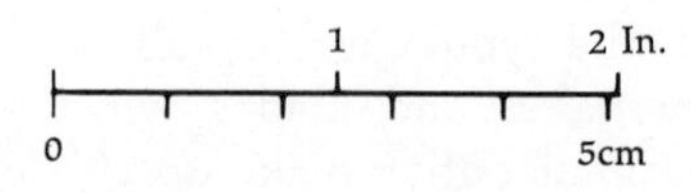

Fig. 19

Pech de l'Azé II.

1, Convergent scraper. 2, Mousterian point. 3, Levallois point core, Layer 4E. 4, Mousterian point, broken. 5 and 6, Crushed flints, Layer 4D.

←

The simple scrapers make up most of the total, transversal scrapers not being numerous (2.9 percent). There was a relatively high percentage of scrapers on the bulbar face, however (2.7 percent). None of the scrapers belongs to the Quina type. Notches (10.2 percent) and denticulates (13.1 percent) are in moderate proportions. There was only one atypical backed knife and no handaxes. The Mousterian points are present but relatively rare (0.7 percent); on the other hand, the Upper Palaeolithic types of tools are well represented for an old Mousterian (group III = 9.3 percent) and among them is a rarity in the Mousterian, a composite tool associating a burin and an end scraper (Fig. 20).

There are 623 flakes and 36 blades that do not show evident traces of utilization, even if most of them may have been used for light duty work, and 1,041 chips; 77 cores are present. A further 367 flakes and tools, heavily bruised, have not been included, as they evidently come from the layer immediately below, which has been cryoturbated.

Thus all the characteristics indicate a Typical Mousterian.

Layer 4B covers layer 4C2, often without any clear sterile layer. The warming up of the climate continues, and now the trees account for 50 percent in the pollens, with a maximum of hazel, and a development of alders and warmth-loving species, among them the oak. In the fauna, the red deer is still dominant (59 percent of the remains, but only 28 percent if the minimum number of represented individuals is calculated), followed by the

→

Fig. 20

Pech de l'Azé II. Typical Mousterian (Layer 4C2).

1, Elongated Mousterian point. 2, Straight side scraper. 3, Convex side scraper. 4, Concave side scraper. 5, Alternate beak. 6, Déjeté *scraper. 7, Burin end scraper. 8, Side scraper, on bulbar face. 9, End scraper. 10, Disk. 11, Denticulate.*

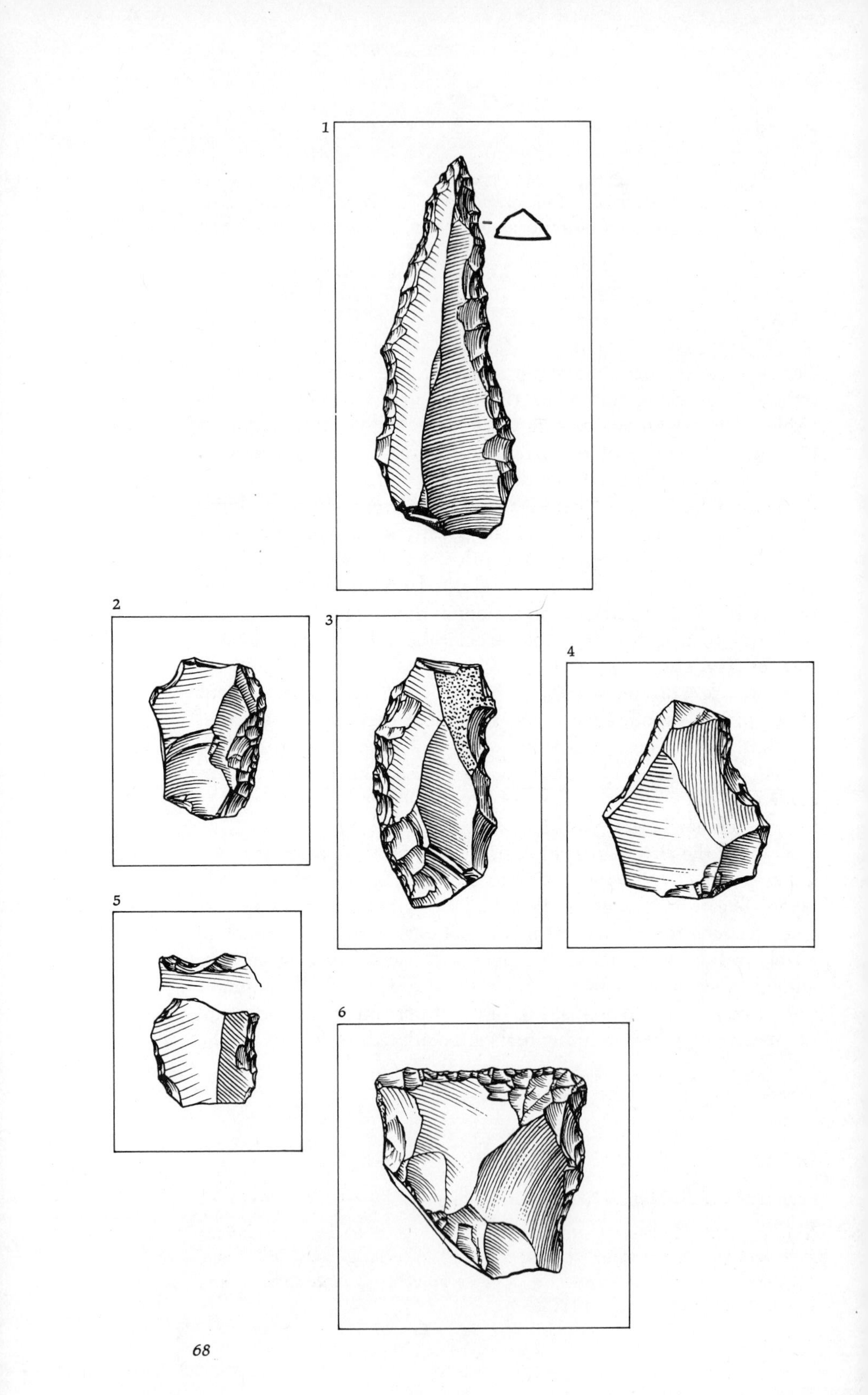
1
2
3
4
5
6

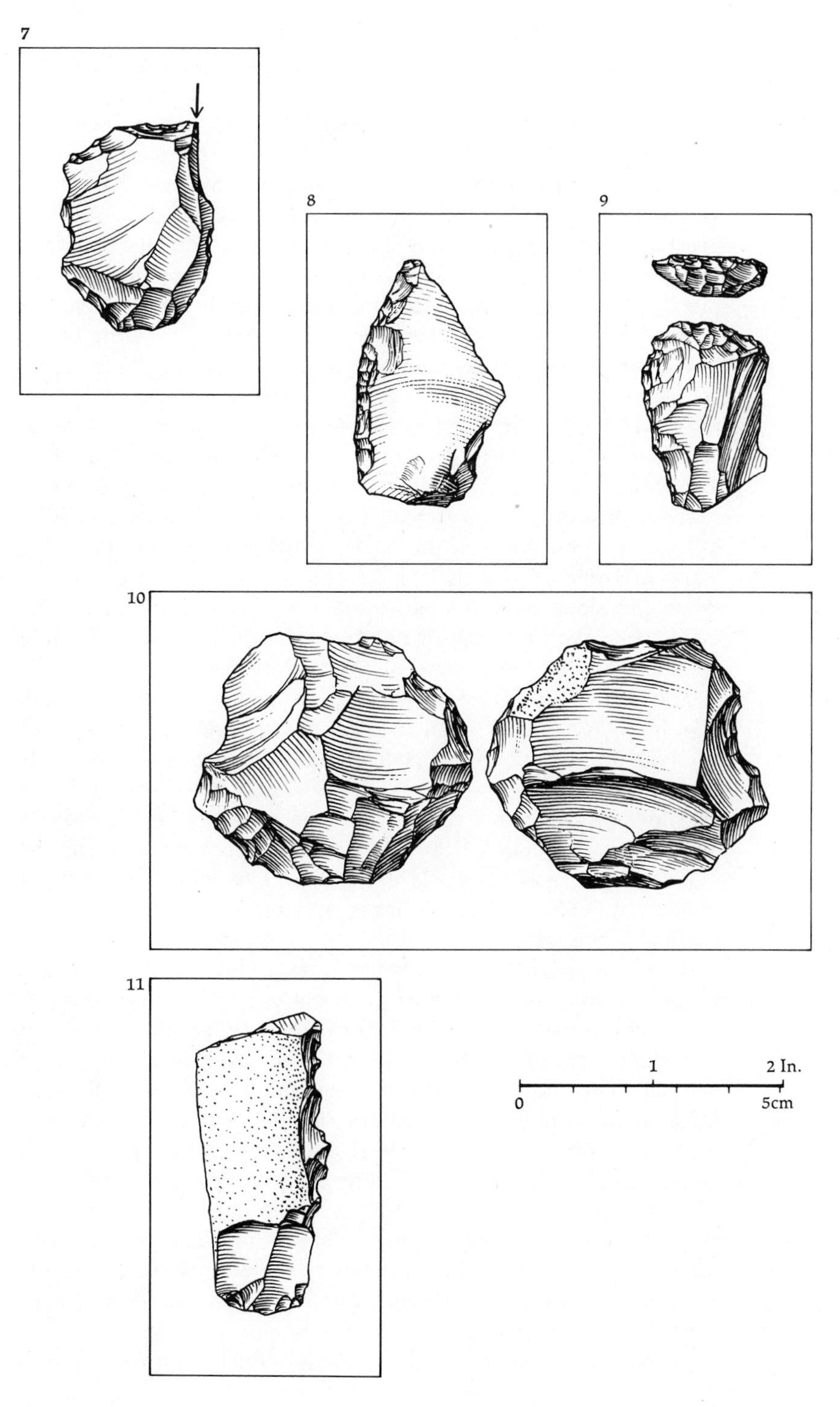
7
8
9
10
11
1
2 In.
0
5cm

horses (28.7 percent of the remains, but 24 percent of the individuals), and the bovids (7.1 percent of the remains, 12 percent of the individuals). Represented by an individual at least is the rhinoceros (and part at least is Merck's rhinoceros, which goes on into the Würm I), the mountain goat, the roe deer, the reindeer, the badger, and the wolf. It is noteworthy that the equids, which made up only 16 percent of the preceding layer, still in a grassy enviroment, are now 24 percent in a forested environment; while the red deer falls from 56 to 28 percent in the maximum number of represented individuals. One would have expected the opposite, and even if the number of remains is not too great (195) and the minimal number of individuals represented even less (23), it looks as if there were a specialized hunting of equids in the Denticulate Mousterian, as the results at Combe-Grenal seem to bear out. This small number of animals also points to something very commonly overlooked, which is that we find only a very small part of the game that has been killed and eaten. Most of the bones were certainly thrown on the slope in front of the shelter, or did not fossilize. Very often, when the slope is excavated, more bones are found in it than inside the cave or shelter. Unfortunately, in the case of Pech II the slope has been destroyed by the railway trench.

The industry of layer 4B belongs to Denticulate Mousterian. The layer was richer than the preceding one, with a total of 6,039 artifacts found. The Levallois index is low (9.4), the faceting indexes low ($FI = 48.9$; $FI^r = 24$), and the laminary index also low (9.4). Except for the FI^r, these indexes do not differ much from the ones of the Typical Mousterian below, but the implements as a whole are larger in size. Typologically, the TyLI is also low (12.5). Utilized flakes and blades account for 18.5 percent of the total.

In essential count, the scraper index is low (14.5), and this value is a maximum, because it is quite possible that some of these scrapers belong to the Typical Mousterian since layer 4B was most of the time stuck on the top of layer 4C2, and this is one case when horizontal excavations would have led to disaster. Most of the scrapers are ordinary side scrapers of sundry subtypes, but among the transversal ones is a Quina type, which gives a Quina index of 0.8. There is one atypical backed knife only, and the Upper Palaeolithic types of tools account for 6.9 percent, most of them being rather poor examples of end scrapers, burins, or borers. Notches are numerous (16.1 percent) and denticulates very numerous (40.3 percent). There are three very atypical handaxes, only partially bifacial (Fig. 21).

There are 1,836 flakes and 159 blades which do not show evident traces of utilization, and 2,382 chips, 175 cores, and a number of basalt and quartz flakes, chips, and pebbles, these last whole or broken, but not hammerstones.

On the whole, a very different assemblage from the preceding one.

Over 4B comes layer 4A, divided into two sublayers, 4A2 (the lower) and 4A1 (the upper). Each contains an occupation level, very poor in both cases, but apparently belonging again to the Typical Mousterian. With 4A2, we have the climatic optimum. The percentage of tree pollens goes up to 57, with a maximum of warmth-loving species (oak, beech, and the rarer hornbeam and maple) and numerous shrubs and ferns. The fauna is scarce—as are the tools—but consists of wild boar, red deer and reindeer, mountain goat, hyena, cave bear, and so on. In layer 4A1, there is a progressive diminution of trees, a diminution of the warmth-loving species, a new development of the hazel, and toward the end, a dominance of pine, willow, and birch. The fauna here seems to contain a higher percentage of reindeer, but the small number of remains makes any conclusion dubious.

Archaeologically speaking, these layers are very poor and show only some very restricted traces of fire. Layer 4A1 has given 179 artifacts; layer 4A2, 337. Most of them are flakes and chips, but the relative percentage of side scrapers and denticulates seems to indicate that they belong to Typical Mousterian rather than Denticulate Mousterian.

With layer 3, the deterioration of the climate continues. It is a cold period, still rather damp at first and then progressively drier. The percentage of tree pollens falls down to 17, then to 8 percent. Some pollens of deciduous trees are still present, birch and mainly willow; but the alders and hazels disappear quickly, and prairies and steppes develop. The fauna does not show a great change. It is still a mixture of temperate types, such as the wild boar and the red deer, and colder types, such as the reindeer or mountain goat. However, the cold types seem to form a higher percentage than previously.

→

Fig. 21

Pech de l'Azé II, Denticulate Mousterian (layer 4B).

1, Levallois flake. 2, Side scraper. 3, Atypical end scraper. 4 to 7, Denticulates. 8, Naturally backed knife. 9, Clactonian notch. 10, Ordinary notch. 11, Tayac point. 12, Atypical handaxe.

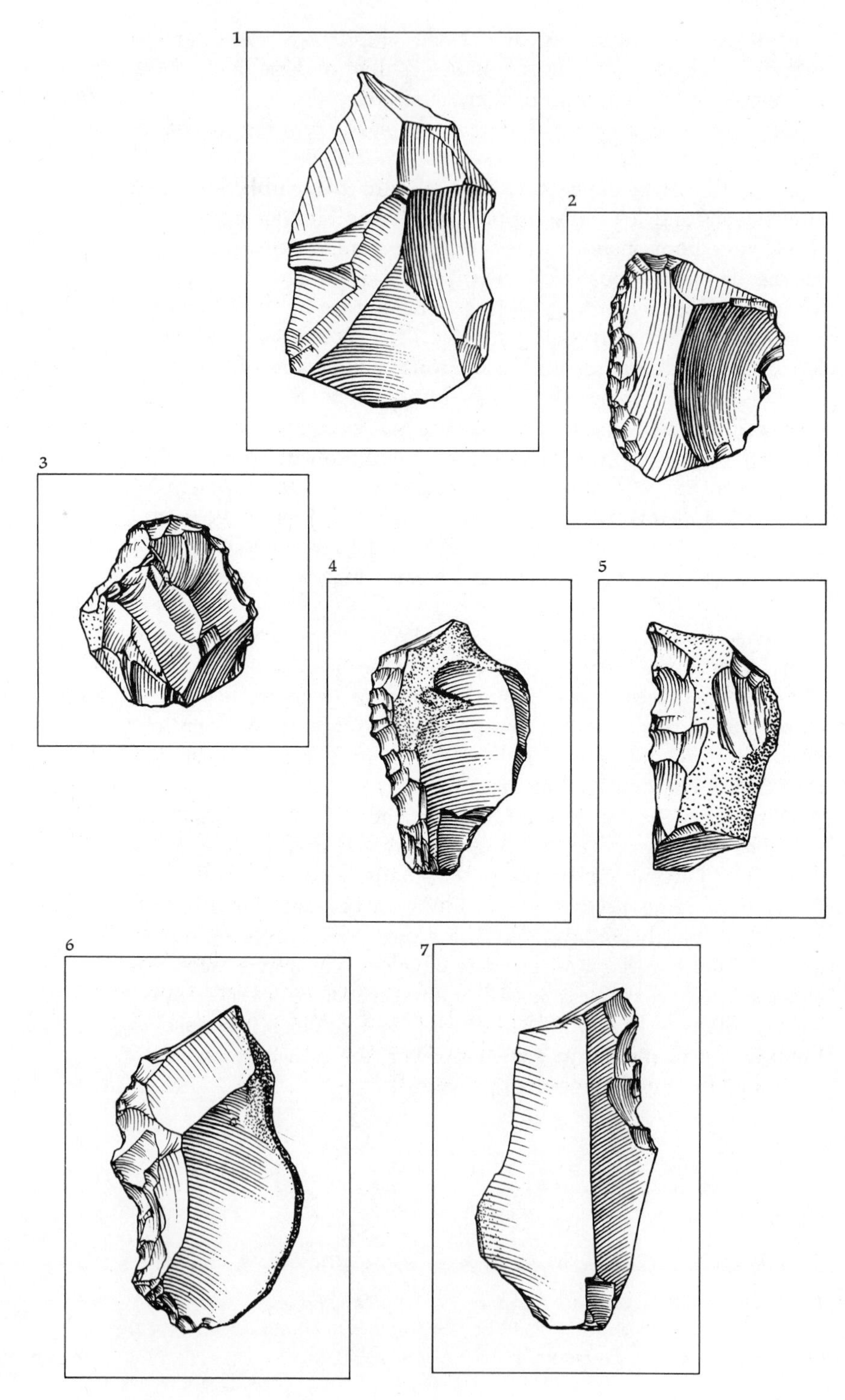
1
2
3
4
5
6
7

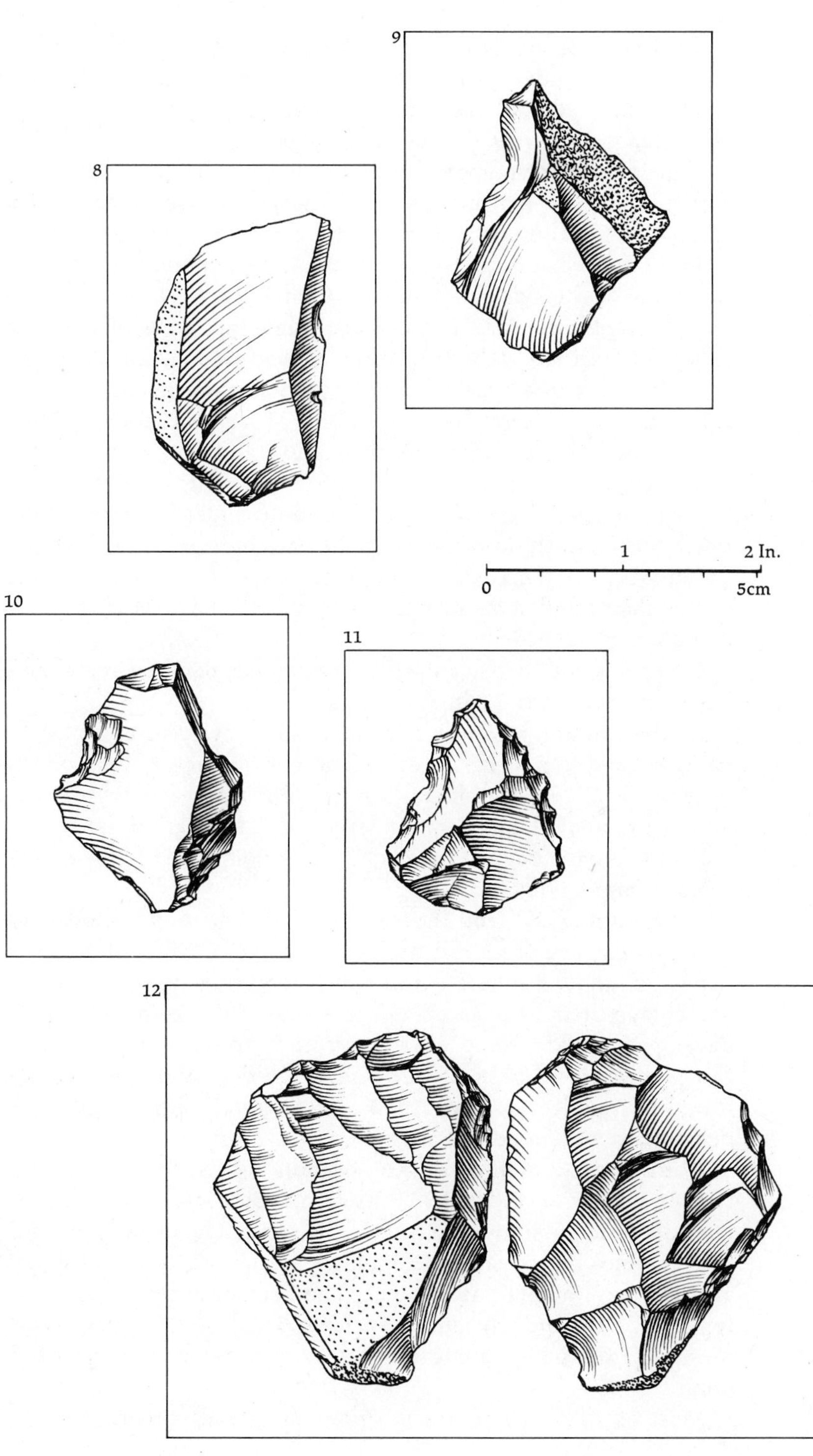
8
9
1
2 In.
0
5cm
10
11
12

This layer is moderately rich, having yielded 1,261 artifacts. The Levallois index is high (29.1); the faceting indexes (respectively, 73.6 and 64.6) are higher than in the preceding levels, but the laminary index (9.3) does not change much.

The Levallois typological index is higher too (23.5) but still rather low. The percentage of ordinary flakes or blades with traces of utilization or accommodation stays around 20 (22.5).

In essential count, the scraper index is of medium value (29.1), and none of the side scrapers is of Quina type (QI = 0). Transversal scrapers are few. The Upper Palaeolithic types of tools are numerous (Group III = 18.2) but a good number of them are truncated flakes, a type which can be naturally made by cryoturbation—and there are indications of cryoturbation in this layer. There is one typical backed knife and one atypical, and two "handaxes" of the nucleiform type. This is not nearly enough to speak of Acheulean tradition; indeed, layer 3 falls much more clearly into the range of variation of Typical Mousterian (Fig. 22). There are 236 flakes, 11 blades, 552 chips, some quartz flakes and some quartz and basalt pebbles, 36 cores, and one piece of red ocher.

Layer 2 is subdivided into different archaeological layers, and has witnessed several climatic changes.

At the bottom, 2G′ was deposited under warmer and mainly damper conditions than layer 3. The percentage of tree pollens goes up to 23, chiefly pine and hazel. Birch and willow are less abundant, and alder, elm, linden, and beech are found sporadically. In the fauna, the red deer is dominant, followed by reindeer and bovids.

At the top of 2G′ and the bottom of 2G, there is a short cold oscillation where trees fall back to 10 percent. Most of 2G was deposited under temperate and damp conditions again; here the percentage of trees goes up to a maximum of 33—hazel is well developed, as are some other deciduous trees, among them a fair percentage of oaks. In the fauna, red deer is dominant, followed by roe deer and wild boar, but a certain amount of reindeer and mountain goat remains.

With layer 2F a climatic deterioration begins, the climate becoming drier and colder. In 2F itself, the percentage of tree pollen falls to 12, mainly pine; in 2E, the percentage is only 10 and some species disappear; in 2D, all the deciduous trees are gone, and the environment is steppic. The sequence continues with layers 2C, 2B, and 2A, and then layer 1, all of them very poor or sterile; the pollen analysis for these layers has not yet been done.

Most of these layers are too poor in faunal remains for the

fauna to have any significance. In the richer 2E reindeer is dominant, followed by bovids.

Not much can be said about the implements found in the different subdivisions of layer 2. Most of them have yielded less than 50 artifacts and very few tools. The exception is 2G, which has given 91 artifacts, among them 27 tools. The high Levallois index (about 20) and the presence of several Quina-type scrapers speaks for a Ferrassie-type Mousterian (Fig. 23). And the simultaneous presence in many of the subdivisions both of typical Levallois flakes and Quina scrapers induces us to think that this type of Mousterian was present all through layer 2. These Mousterians seem to have used this end of the cave for short stays only; most of the occupation must have been on the other side. Unhappily, no traces of this layer are preserved in Pech I, but in the rear part of cave I a good number of Quina scrapers have been found mixed with other material.

During the time of deposition of layers 2G′ and 2G, the climate was not too cold, but afterwards it deteriorated fast. The explanation may be that under cold conditions Pech I, better oriented, would have been a more pleasant abode than Pech II, which was open to west winds.

PALETHNOLOGICAL OBSERVATIONS ON THE MOUSTERIAN AT PECH DE L'AZE II

Most of the work has yet to be done here. The analysis of the notebooks, interpretation of the maps of the layers, and so on, is a lengthy task, so that we have to rely on the observations made on the spot during the excavations.

From the fireplaces and the position of the tools, it is clear that the Typical Mousterians of layer 4C2 camped mainly under what is now the shelter, while the Denticulate Mousterians of layer 4B lived mainly in the entrance of the cave. This was why, when we made a first test pit in the entrance, we thought we had found only Denticulate Mousterian. The Typical Mousterian

→

Fig. 22

Pech de l'Azé II, Typical Mousterian (layer 3).

1, 2, 3, Levallois flakes and blade. 4, 5, Convex side scrapers. 6, Straight side scraper. 7, Concave side scraper. 8, Déjeté *scraper. 9, Convergent scraper. 10, Denticulate. 11, Small end scraper. 12, Borer. 13, Notch.*

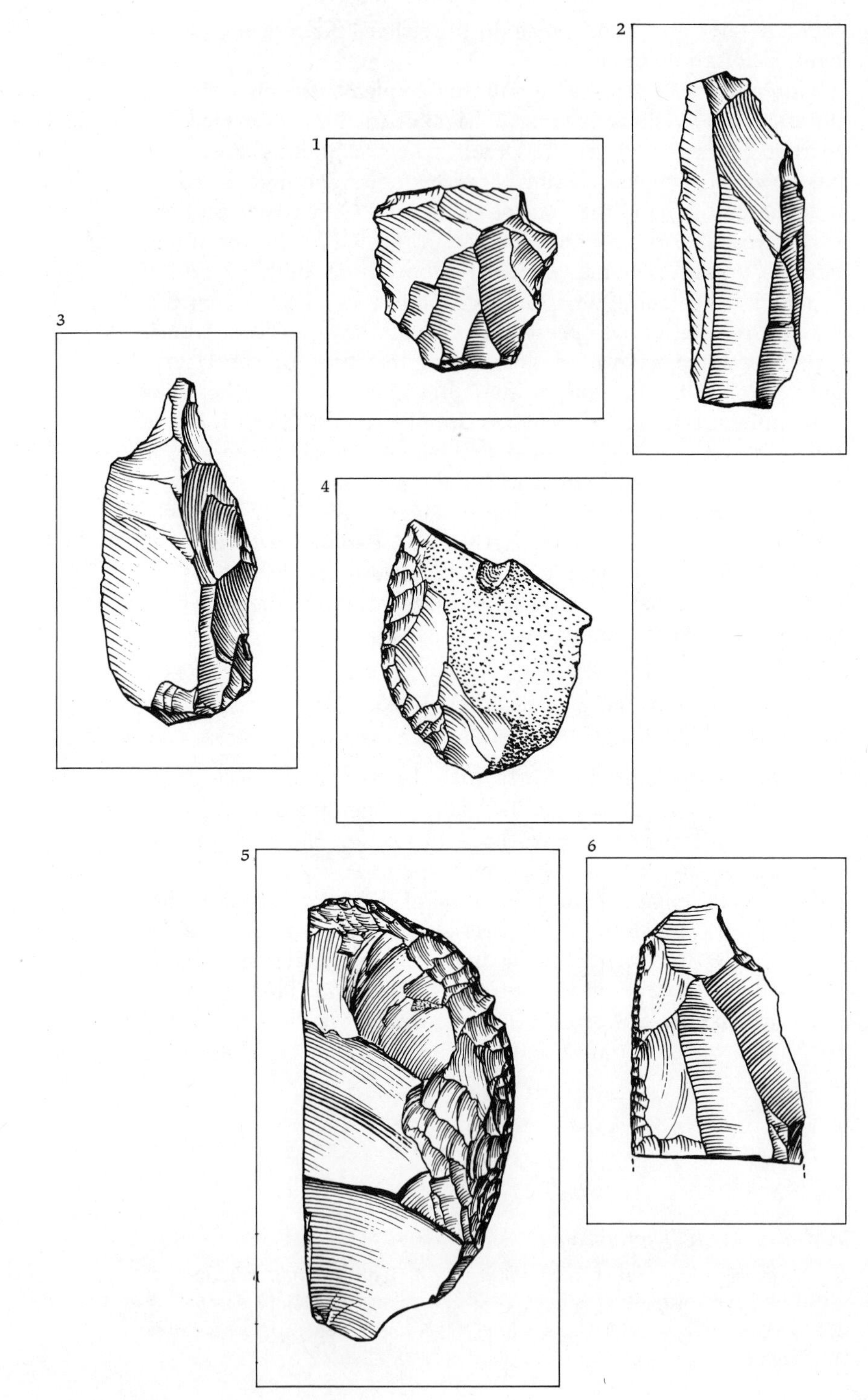
1
2
3
4
5
6

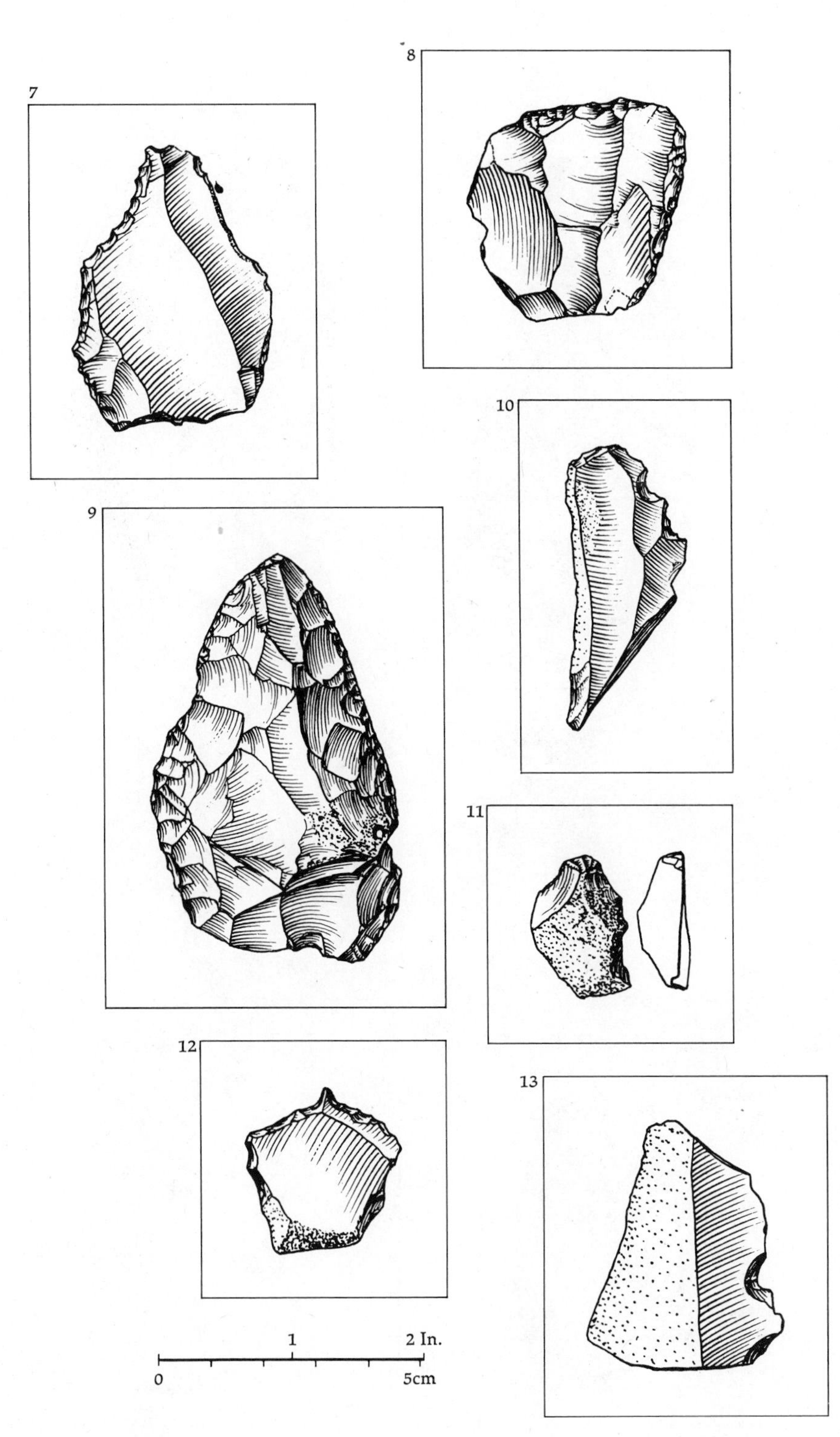
7
8
10
9
11
12
13
1
2 In.
0
5cm

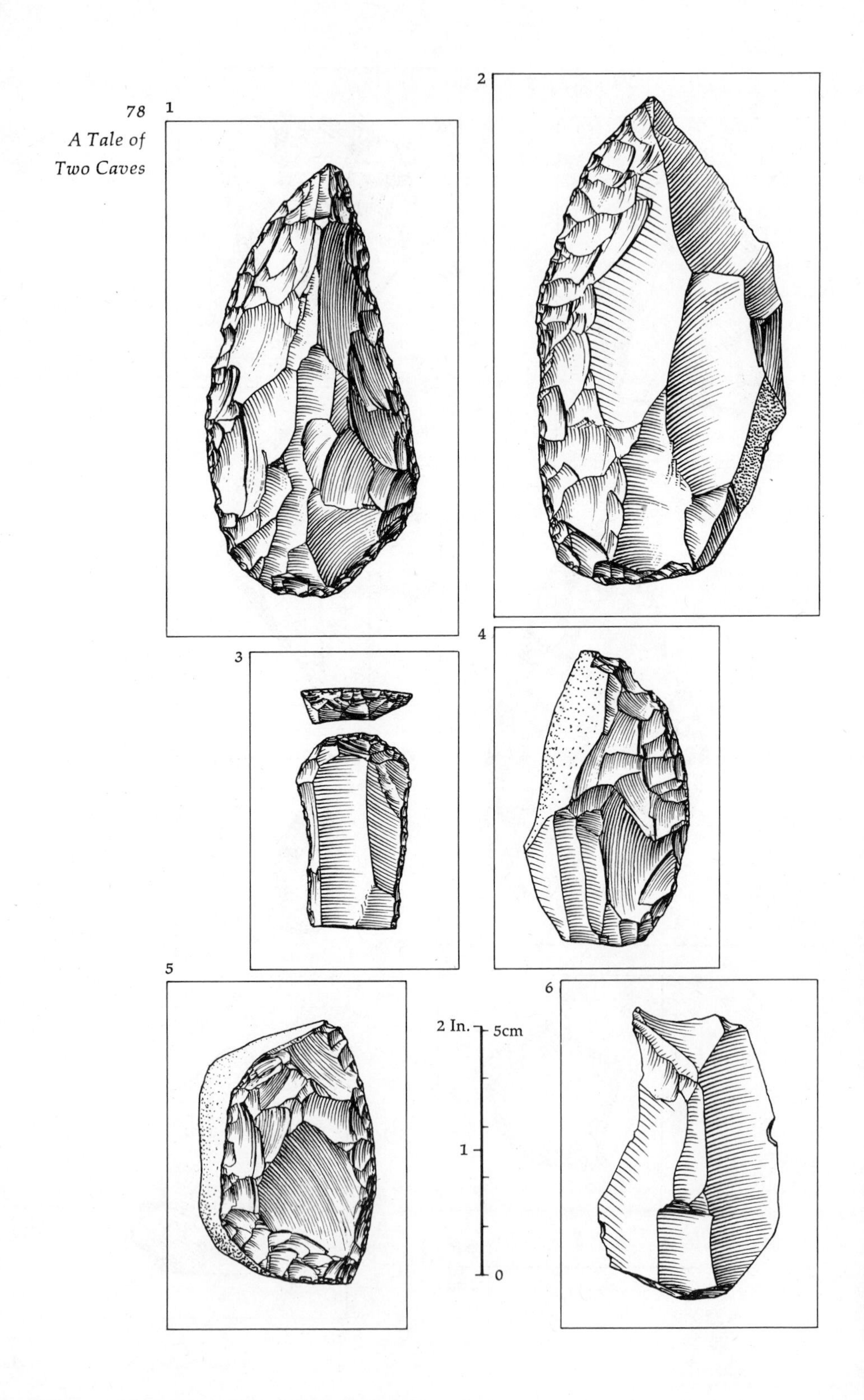
1
2
3
4
5
6
2 In.
5cm
1
0

component of this mixed assemblage was too small, but the reasons for this different choice of habitat are unknown.

At different levels (mainly however in 4B), we have found clusters of big limestone blocks, clearly assembled by man, but representing neither a fireplace (no ashes, no reddening of the stones) nor, as we had hoped at first, a sepulcher.

Inside the cave, it is not possible to separate layers 4A1 and 4A2. The composite layer is just called 4A. And it is here, in square F12, that a very interesting object was found: a pierced bone. In Upper Palaeolithic, this would be nothing new, but in a Mousterian context it is surprising. A hole has been bored into the bone of an animal of red deer size at least; the object is incomplete, and we have only about two-thirds of the circumference of the hole (Fig. 17). We do not know what such an implement was used for: was it an ornament, or did it have a more utilitarian purpose, as a thong stropper, for example? We know of no other occurrence of such an artifact in the Mousterian (9).

THE PALEOLITHIC SEQUENCE AT PECH DE L'AZÉ I

In our 1949–1951 excavations we had established, as I said before, the general stratigraphy of the site, a stratigraphy which was good enough for our purpose at the time but should now be refined. This was why, in 1970, we began renewed excavations at Pech I and at the same time also on Pech de l'Azé IV, a collapsed shelter situated about 100 meters eastward. However, as I write this, the results of the new excavations at Pech I bear mainly on the topmost layers, close to the wall of the shelter, which had not been touched in the 1949 excavations and which are largely sterile. So, we shall use the results of our old excavations here, with some additions and corrections.

As explained earlier, nothing is left *in situ* inside the cave, with the exception of the breccias. Just under the entrance of the

←
Fig. 23

Pech de l'Azé II.

1, Convergent scraper (layer 2E). 2, Convex side scraper, Quina-type (layer 2G). 3, Upper Palaeolithic end scraper, found on the surface inside the cave. 4, Convex side scraper, Quina-type (layer 2E). 5, Convex side scraper, Quina-type (layer 2F). 6, Levallois flake (layer 2E), Quina—Ferrassie Mousterian.

cave are two parallel dry-stone walls, one containing Peyrony's back dirt, the other one Vaufrey's. It was in the small part that they excavated that Capitan and Peyrony found the child's skull in 1909, "approximately under the overhang and close to the wall of the cave," as Denis Peyrony told me some years before his death. We shall see that the exact stratigraphical origin of this skull is open to question.

In the part of the site that remains today, the top layers are practically sterile (Fig. 6). They are made up of large and small blocks, corresponding to a rapid collapse of the roof of the shelter which prolonged the cave. Under them are other fragments, also both large and small, which contain, from top to bottom, archaeological layers 7, 6, and 5. Layers 7 and 6 are in fact composed of several sublayers, more or less scattered among the blocks. At that time, the real occupation zone was probably deeper inside the cave but has been destroyed by former excavations. Layer 5, not very rich in the front part, rests on big slabs of limestone, representing a heavy roof fall. All these layers belong to the Mousterian of Acheulean tradition, type B.

Under the slabs is the very rich layer 4, sandy and blackish in color. Its thickness varies from place to place between 1 and 10 inches. It contains Mousterian of Acheulean tradition, type A. Below it are some inches of sands, then layer 3 (the same kind of Mousterian), then sands again (layer 2), sterile, then stratified sands (layer 1).

In all these layers, the fauna is relatively rare, but reindeer, albeit rare, was present at different levels. The pollen analyses have been made at present only for layers 4 and 5. Layer 4 was deposited under very cold and damp conditions, with only about 5 percent tree pollens, mostly pines. The climate of layer 5 was also very cold, but drier.

Of course, new analyses will be made, for all the layers; that is one of the reasons we took up this excavation again.

The Lower Layers

We published layers 3 and 4 separately in our 1954 excavation report (4), but as there is very little change and little, if any, difference between them, here we will describe layer 4 only, since it is by far the richer (Figs. 24 and 25).

Layer 4 has given a total of 35,100 artifacts. The Levallois index is low (8.4), but high enough to be sure that these Levallois flakes are not accidental. In the case of the Mousterian of Acheulean tradition, special care must be taken not to mistake the thin flakes obtained in making handaxes for true Levallois flakes. Both show a well-prepared upper surface (in the handaxe flakes, this comes from preceding work on the tool), but the characteristics of the butts of the flakes are different. Levallois flakes show a normal butt, and most of them being struck out with a hard hammer, they have well-developed conchoids. Handaxe flakes, on the other hand, were generally struck out with a soft hammer (wood, antler, or bone) and so show an extremely distinctive butt, very narrow, with a kind of overhang over the ventral face (Fig. 26) and a very diffuse bulb. These handaxe flakes are not counted in the technical study, since they are, theoretically at least, waste flakes, even if many of them have been used to make scrapers or other tools.

The majority of the butts are faceted (FI = 57.3), even if the number with small facets is only a little over one-third (FI^r = 36.7). The laminary index is rather low (11.1).

In real count, the typological Levallois index is low (4.5) and used flakes account for about 31 percent. In essential count, the scraper index is of medium value (SI = 35.2), but the scrapers are very diversified, the most numerous being the straight and convex side scrapers. Transverse scrapers are few (about 1 percent), but one of the characteristics of this assemblage is the relatively high percentage of scrapers made on the bulbar face (6.1 percent), of bifacially worked scrapers (1.1 percent), and of double scrapers with alternate retouch (2.2 percent). Points are rather rare (0.9 percent), but some of them show a careful thinning of the butt, probably to facilitate hafting. The percentage of Upper Palaeolithic types of tools is not specially high (10.7 percent), but these tools are well made and often typical. One tool, usually scarce in other types of Mousterians, is very well represented here—the Mousterian "raclette" (6.5 percent). This is made on thin handaxe flakes, retouched by a small nibbling retouch, often alternate; some are almost microlithic (Figs. 24 and 25). Notches (10.5) and denticulates (18.2) are numerous. There is one foliate bifacial point—although not completely bifacial, it is true—which is a rarity in French Mousterian of any kind.

⟶

Fig. 24

Pech de l'Azé I. Mousterian of Acheulean tradition, type A (layer 4).

1, 3, Side scrapers on bulbar face. 2, Double side scraper on a blade. 4, Convex side scraper. 5, Denticulate. 6, "Raclette" on a bladelet. 7, Small borer. 8, Déjeté *scraper. 9, Mousterian point with partial bifacial retouch. 10, Nosed end scraper.*

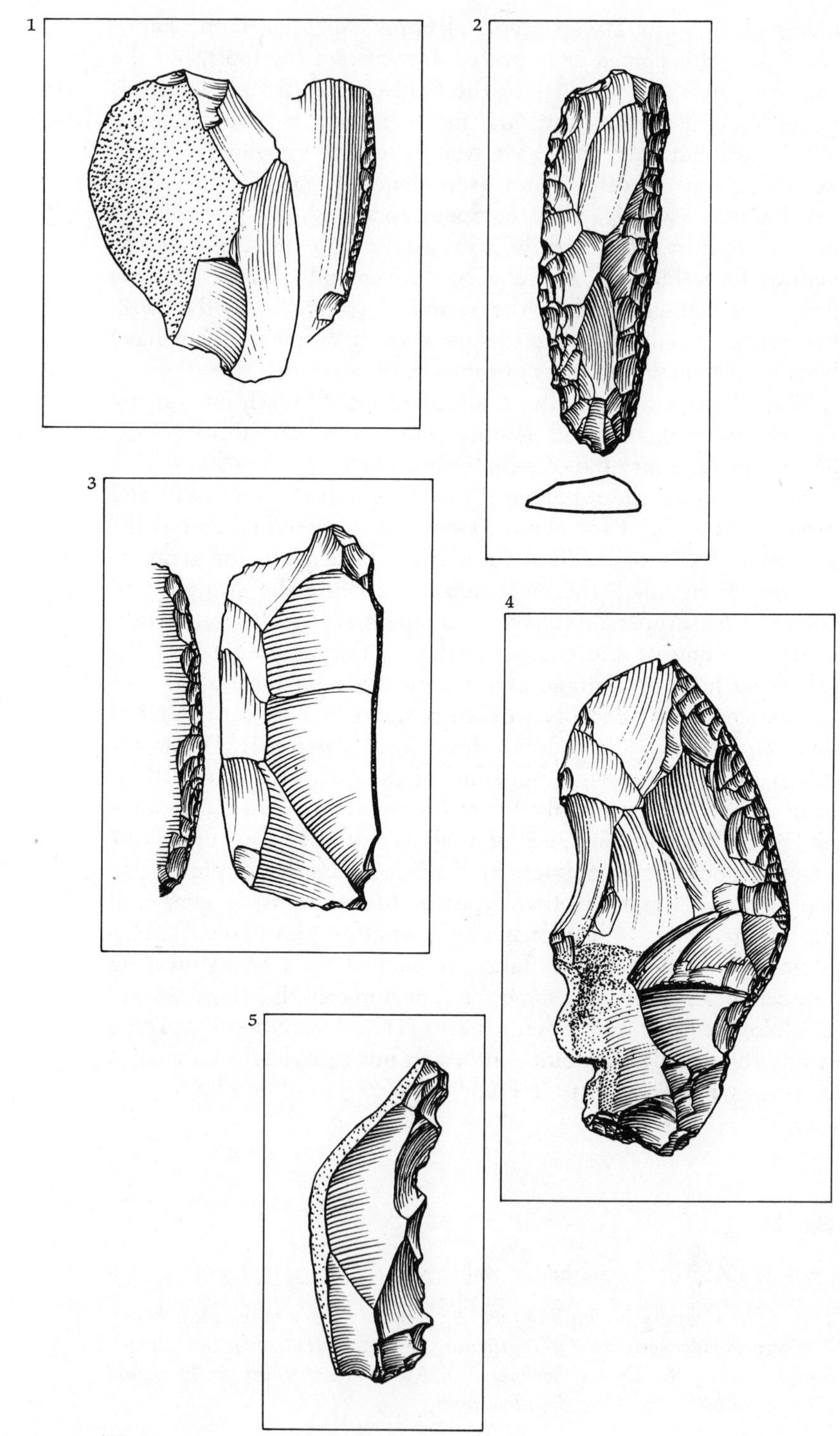
1
2
3
4
5

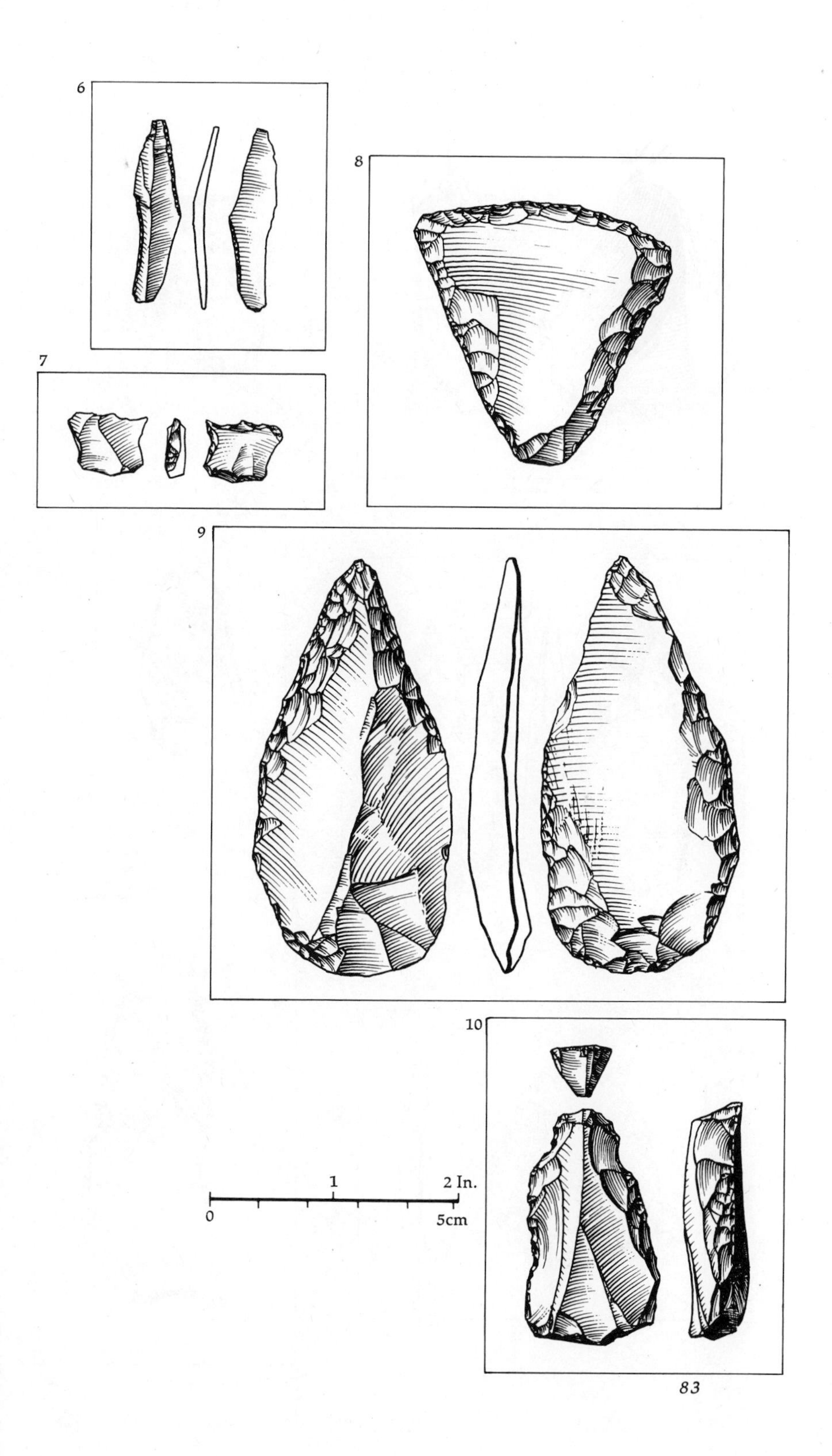
6
7
8
9
10
1
2 In.
0
5cm

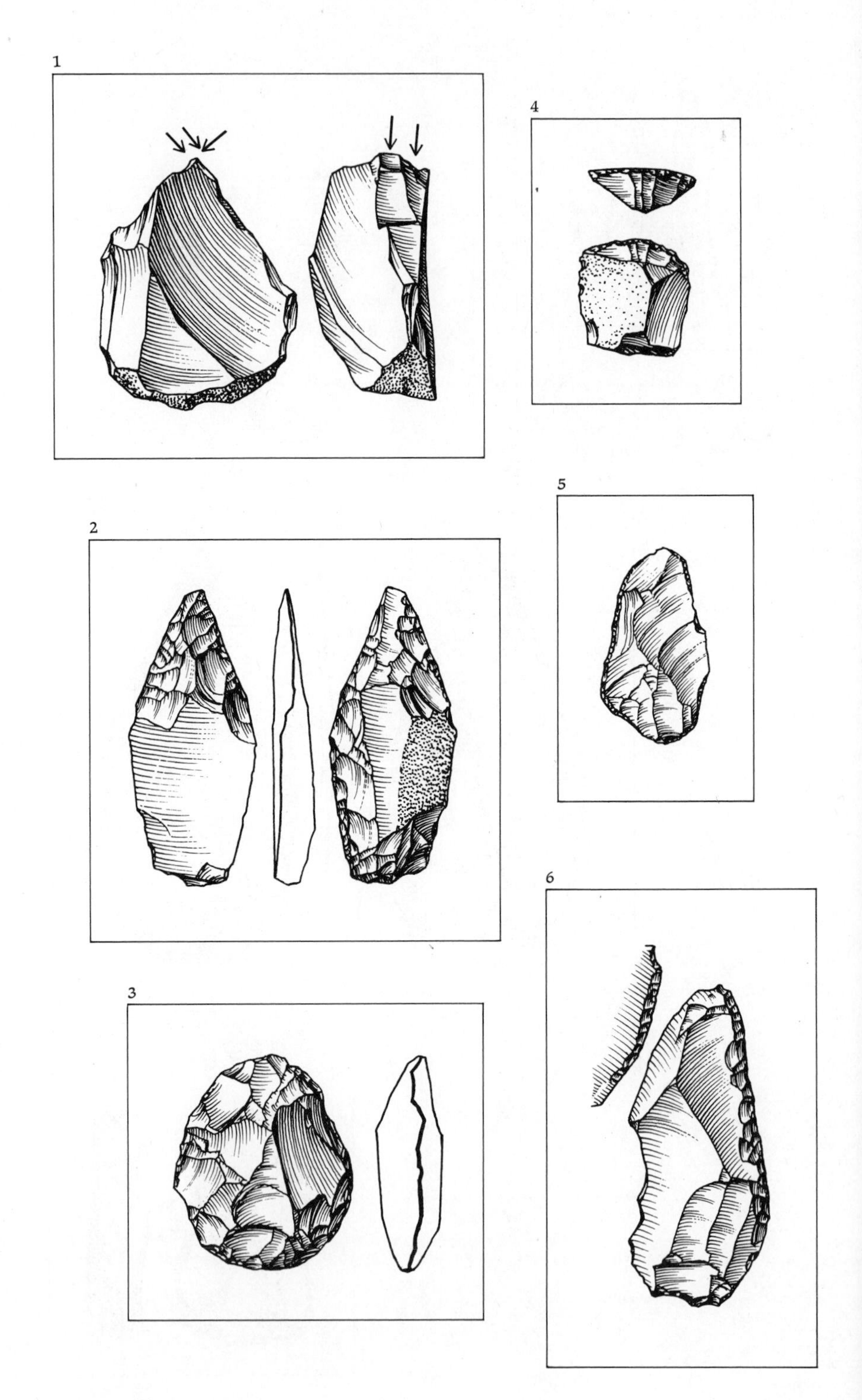
1
2
3
4
5
6

7

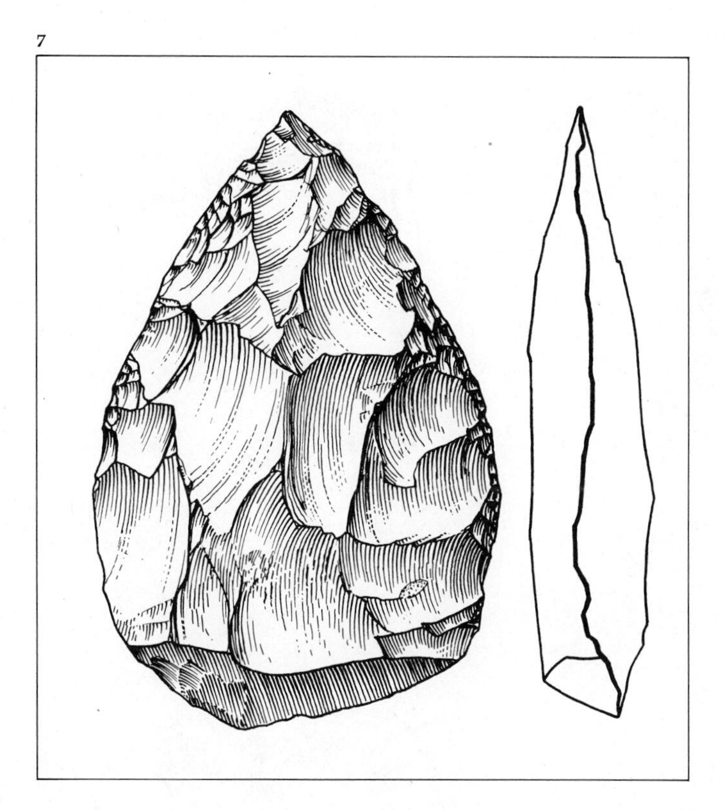

8

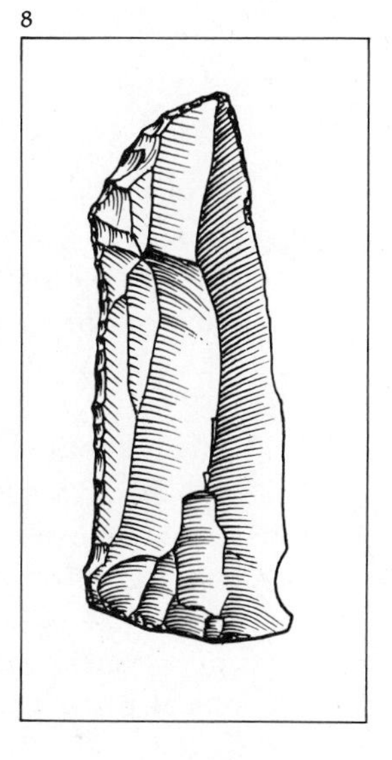

1 2 In.
0 5cm

Fig. 25

Pech de l'Azé I, Mousterian of Acheulean tradition, type A (layer 4).

1, Burin. 2, Bifacial point. 3, Small discoïdal handaxe. 4, Small end scraper. 5, "Raclette." 6, Alternate double side scraper. 7, Cordiform handaxe. 8, Backed knife.

1

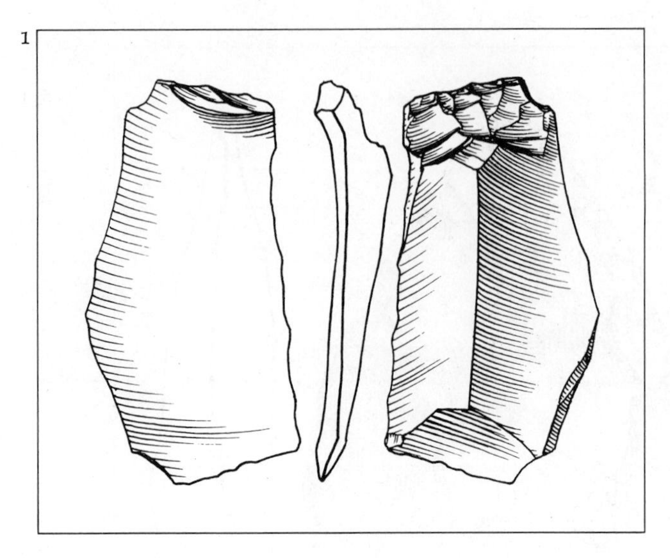

2

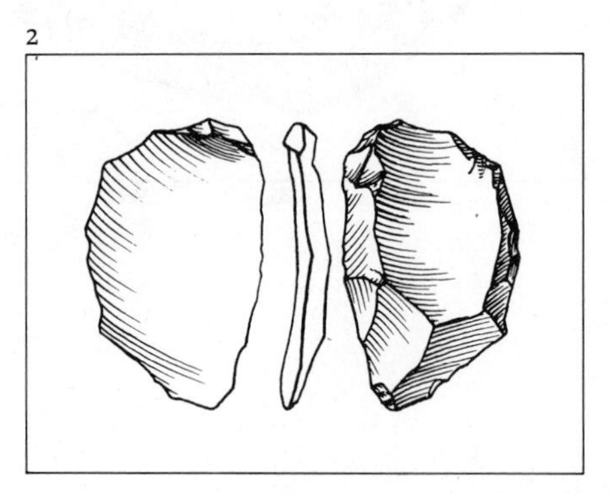

Fig. 26

Handaxe flakes, Pech de l'Azé I, layer 4.

Notice the narrow shape of the butt and its obliquity over the ventral face of the flake. This butt often shows a kind of overhang over the ventral face.

The handaxes make up 5.8 percent of the total of tools, and number 156, counting the fragments. There are some flat, triangular handaxes, but the most common is the cordiform handaxe and its subtypes. One other important type is the "partial handaxe," that is, a tool which has the shape and all the general characteristics of a handaxe but is only partially retouched on one of its faces.

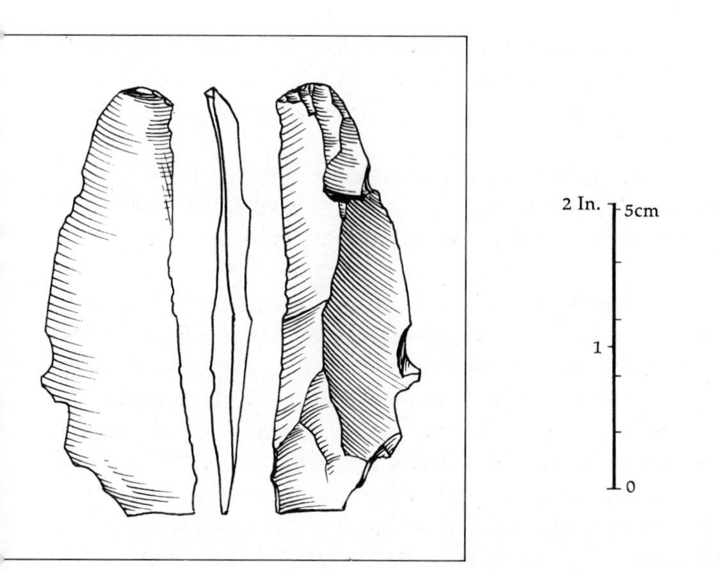

4

There are 3,522 flakes, 314 blades, 24,875 handaxe flakes, 700 chips, many basalt and quartz flakes and pebbles, and 278 cores to complete the assemblage. There are also a lot of mineral pigments.

This assemblage corresponds very well to the definition of Mousterian of Acheulean tradition, type A.

The Upper Layers

Layer A, which seems the oldest of the upper layers, is special in that it appears to represent a transition between types A and B of Mousterian of Acheulean tradition. Unhappily, it is also poor;

only 347 artifacts have been found in it. The Levallois index is higher (13.5), as well as the laminary index (13.1), but the indexes of faceting are lower than in layer 4, being 54.1 and 27, respectively.

The scraper index does not change much. The pseudo-Levallois points, which will be a characteristic of the upper layers, begin their development here and rise from 1.5 in layer 4 to 7.2.

The upper layers proper show an evolution—with a general lowering of the Levallois index—of the faceting indexes and a general augmentation of the laminary index. The scraper index falls down considerably, to a value of about 4 in layers 6 and 7. One interesting feature is the development of the pseudo-Levallois points: these are triangular (when typical) or pentagonal flakes, obtained in a special way from discoidal cores. They are seldom retouched, and it is debatable whether they are true tools, as a Levallois point, or only a waste.* While they play a very minor role in layer 4, they begin their development in layer A and go up to a percentage of about 16 in layers 6 and 7.

The percentage of backed knives rises to 11.7 in layer 7, and some of them are already almost of the Châtelperron type which will characterize the first Upper Palaeolithic culture in France—the Lower Perigordian (also, mainly by English-speaking people, termed the Châtelperronian). The percentage of other Upper Palaeolithic type of tools (end scrapers, burins, borers, truncated blades, and flakes) goes up too, and, what may be even more significant, they are now often very typical (see Fig. 27). The percentage of handaxes decreases to a low of 2.2 in layer 6; they are usually small and not too well made. Denticulate tools are numerous and show a tendency to increase in proportion. These will still play an important role in the Lower Perigordian.

Looking at the assemblages of these upper layers of Pech de l'Azé I, one gets the impression of something no longer purely Mousterian, but not yet Upper Palaeolithic. This evolution does not continue at Pech I, since the uppermost layers, representing the end of Würm II, are almost sterile; but it probably went on in other places, and its outcome may well be the Lower Perigordian.

* They are called pseudo-Levallois points because they are triangular flakes obtained in this shape without retouching, like the Levallois points. But the axis of the flake is oblique to the axis of the triangle, instead of more or less coinciding with it as in the Levallois points (see Fig. 27.5).

It should be recalled here that we have no certain human remains from any Mousterian of Acheulean tradition layer. It has often been said that the skeletons of Spy (Belgium) and Le Moustier belonged to this type of Mousterian, but it is easy to demonstrate that they belong, in fact, to other industries. This leaves the child's skull found at Pech de l'Azé I, in 1909, by Capitan and Peyrony. It is the skull of a young child, less than four years old, which has recently been studied again by several authors (20). It belongs to the Neanderthal race, with some modern characteristics. But does it really fit in with the Mousterian of Acheulean tradition?

Capitan and Peyrony do not give many details of its discovery, unfortunately, but the few they do give do not fit such an attribution. They say:

> There exists in that cave an archaeological layer, formerly excavated in an unscientific manner. This layer continues on the terrace that lies in front of the cave. There, it is about 1 meter thick, but is covered with 3 meters of big limestone blocks and debris, resulting from the collapse of the ceiling, which jutted out in front of the cave as a kind of shelter.
>
> When we had removed the debris completely, the top of the archaeological layer was found absolutely intact. Ten centimeters deep, inside the layer, we found the skull of a child, five or six years old, crushed. All around it lay numerous artificially broken bones, and the teeth of bovids, red deer, horses, goats, and reindeer, then many flints: points and knives—scrapers well worked on one face only, of Upper Mousterian type. Below the skull, the Mousterian layer contained fine axes of the Saint-Acheul type (17).

⟶

Fig. 27

Pech de l'Azé I. Mousterian of Acheulean tradition, type B (upper layers).

1, Backed knife on a flake. 2, Denticulate. 3, Double burin. 4, Convex side scraper. 5, Pseudo-Levallois point. 6, Backed and truncated flake. 7, Backed knife on a blade. 8, Backed knife on a blade very similar to the Châtelperron-type. 9, Blade. 10, End scraper. 11, Small triangular handaxe. 12, Prismatic core for bladelets.

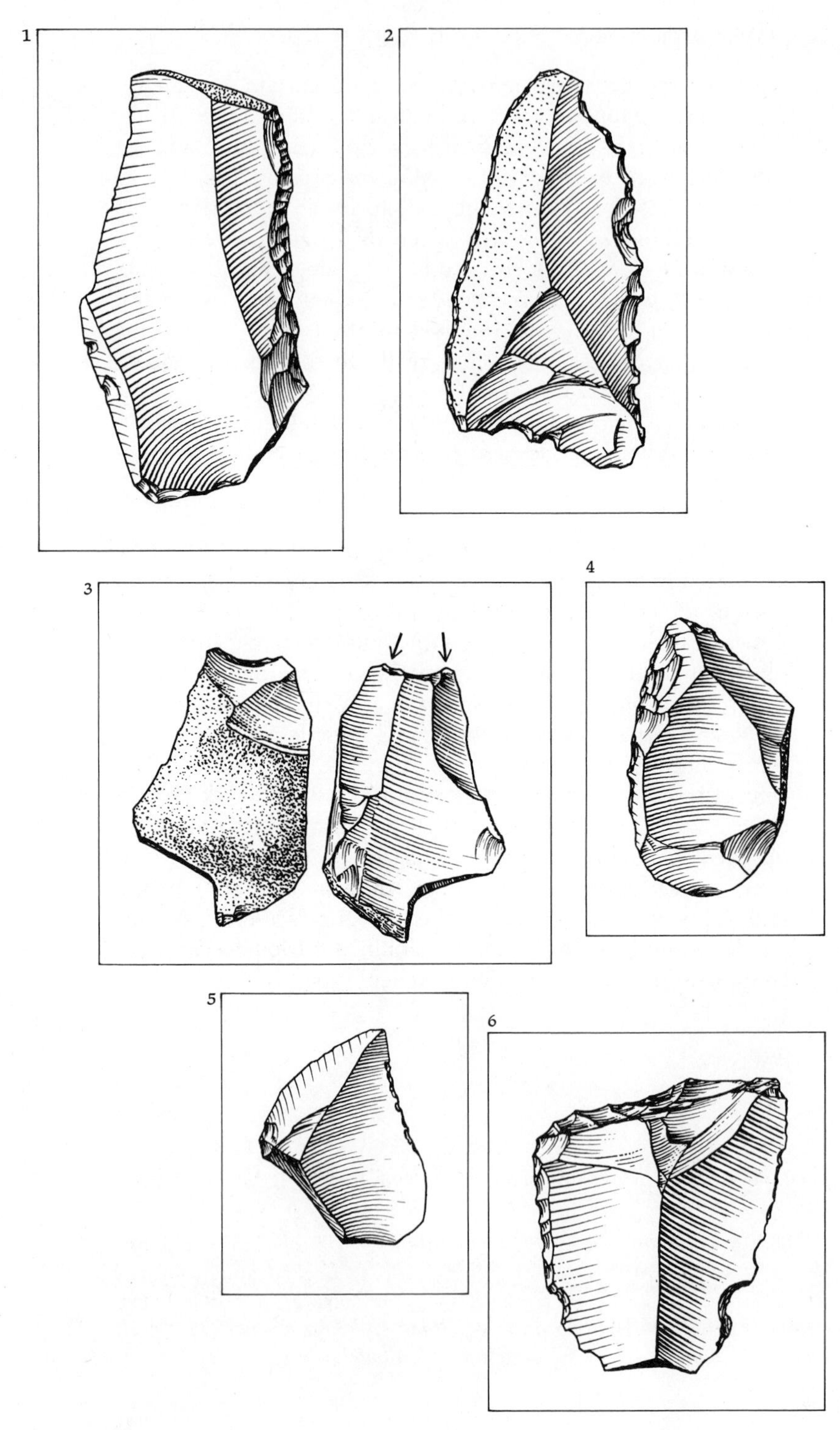
1
2
3
4
5
6

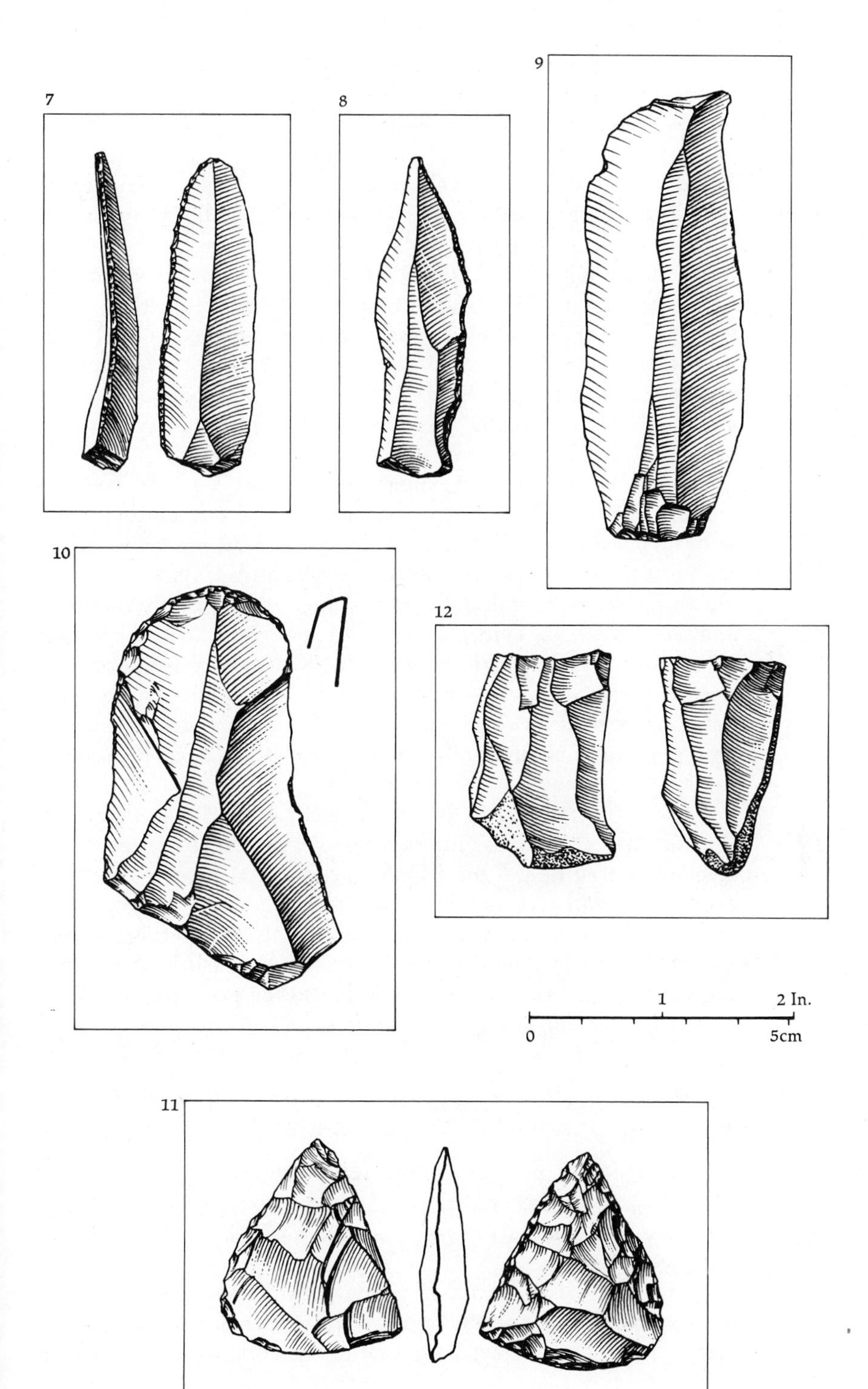
7
8
9
10
12
11
1
2 In.
0
5cm

From this description, it is clear that the skull does not belong to our layer 4, rich in handaxes. Could it belong to layers 5 or 6? It does not seem likely. Reindeer is very scarce in these layers, if present at all. And nobody would call the points and scrapers found in these levels "well worked . . . of Upper Mousterian type"; they are usually rather "decadent." On the other hand, we should remember that at the time of Capitan and Peyrony's excavations, what was called "Upper Mousterian" was Quina-type Mousterian. The evolution commonly admitted then was: Mousterian of Acheulean tradition, the handaxes being an Acheulean "hangover"; then Typical Mousterian; then "Upper" Quina Mousterian. It is a great pity that the tools found with the skull are lost. The fauna described seem to fit better with those of layer 2, in Pech de l'Azé II, than with those of Pech I, where reindeer is always rare, and sometimes absent.

Furthermore, a fluorine test has been carried out, although not published, on the skull, and it seems to be about ten times richer in fluorine than bones from any of the Würm II layers.

All this seems to indicate that the skull was probably found in another remain of Würm I deposits, analogous to the breccias and under which the Mousterian of Acheulean tradition extended. After all, at the beginning of the century, everything here was termed "Mousterian." The Mousterian was considered an entity, and nobody really bothered about stratigraphical niceties for that period.

Another hypothesis is that the skull really belongs to Würm II, and that there was, interstratified with the Mousterian of Acheulean tradition, a brief Quina occupation. Perhaps the new excavations will give us the final answer.

Why are human remains unknown in the Mousterian Acheulean tradition? It is hard to answer this, but possibly these people had different funeral habits; perhaps they exposed their dead on trees or platforms outside the caves.

PALETHNOLOGICAL OBSERVATIONS ON THE MOUSTERIAN OF ACHEULEAN TRADITION AT PECH DE L'AZÉ I

As we have said, no real attempt was made in the 1949 excavations to take into account the horizontal disposition of the tools, since the excavated area was so small. However, some observations were made.

Under layer 3 is a kind of pavement of limestone slabs, but it is impossible at present to tell if this is an intentional pavement, man-made, or just the result of the collapse of the ceiling of the shelter. The new excavations may give us an answer here.

An interesting feature was the dry-stone wall, built in the extension of the south side of the cave. It was about 1 foot high and 1½ feet wide at the base, and contained layer 4, which existed inside the wall but not outside. Unfortunately, we could follow this wall only in the small part we excavated; beyond, if it existed, it had been destroyed by previous excavations.

In layer 4 we found a great quantity of mineral pigments: 103 amorphous blocks of black manganese dioxide, and 67 which showed traces of utilization. Some were rectangular or triangular "pencils"; others had been scratched with flint tools to get a black powder that can be smeared with the fingers. Others again were rounded, with a smooth surface showing they had been used to trace lines or spots on a not-too-hard surface: animal skins, or just human skins? If they were used for body paint, since they give a blue-black color, one could deduce that the Mousterian's skin was rather light-colored. Three fragments of red ocher only have been found, and some traces of yellow ocher, but this does not necessarily mean that the Mousterians had a preference for black: manganese dioxide is abundant in the country, ocher less so (Fig. 28).

One flat stone, of gritty limestone (measuring 11 by 9.5 by 3 cm), seems to have been used as a palette to grind black color (Fig. 28.5).

There were two centers of localization for the scrapers: one close to the dry-stone wall, near the entrance of the cave; the other in the closest part to the rock wall of the shelter that had then been excavated. It seems that concentrations of handaxes also existed. There was definitely a big concentration of handaxe flakes in the spot indicated on Figure 6, as if this had been a workshop for handaxes.

From the study of faunal remains, Jean Bouchud, who examined the fauna of the first excavation at Pech I, could deduce that, during the deposition of layer 4, the cave was occupied at least from May to October. In the upper levels the cave was occupied all the year round, but the material for the summer months is more abundant than that corresponding to the winter months.

BACK TO THE BRECCIAS INSIDE THE CAVE

As I said, we first believed these breccias to be the remains of Würm II sediments, left by preceding excavators because their hardness and awkward position made excavation difficult. When,

Fig. 28

Pech de l'Azé I, Mousterian of Acheulean tradition (layer 4).

1, Manganse dioxide block with incisions. 2, Smooth rounded block. 3, Triangular pencil. 4, 6, Scratched blocks. 5, Palette.

one rainy day, we decided to test them, we expected to find in them Mousterian of Acheulean tradition. As we have seen, what we found was different, and rather puzzling, until we excavated Pech II.

There were—and in some parts there still remain—two main "suspended breccias" along the right side of the cave (looking

5

6

inside) (Fig. 5). They were so hard that we were obliged to use hammer and chisel, and of course, this kind of excavation does not give a particularly accurate stratigraphy. Also, no different layers were visible. Everything was a light buff color, which is the color of the sands of Pech de l'Azé when unweathered. Close to the wall there were no traces of fire, or beds of debris, to guide us. So, we took arbitrary limits, and excavated the thicker one in three artificial layers, and the other thinner one in two parts. Of course, there is no guarantee at all that the arbitrary limits we chose were the true limits between archaeological levels. We also found, in the old dumps inside the cave, two isolated blocks of breccia, which we excavated without having any way of knowing which side had been up and which down.

Breccia I (closest to the entrance of the cave), then, was divided arbitrarily into two layers. The bottom part gave a scraper index of 14.9 and a denticulate index of 28.1, on 225 artifacts (128 tools). This could be Denticulate Mousterian. The Levallois index was 7.7, which could fit well enough with layer 4B in Pech II, except that the percentage of denticulates is much lower; but in 1952–1953, the same layer 4B, outside the Pech II cave, had given a denticulate percentage of only 35. The top part of breccia I was very poor (46 artifacts, 19 tools), so the indexes here are meaningless.

Breccia II was subdivided into three parts. The lowest yielded only 40 artifacts (24 tools) and not much can be said, except that, in this small sample, scrapers and denticulates have the same proportion (25 percent). The middle part is richer (182 artifacts, 105 tools). Its Levallois index is 15.1 and its scraper index 22.8, while the percentage of denticulates is 19.0. This could correspond to Typical Mousterian with a mixture of Denticulate Mousterian. The upper part, unfortunately poor (52 artifacts, 38 tools), gives, with all the reservations so small a sample implies, a scraper index of only 13.1 and a very high percentage of denticulates—52.6. This seems to be Denticulate Mousterian.

We have tried some pollen analysis on breccia I. The cave bear sands, at the extreme bottom, have given indications of a very cold climate which could correspond to Riss III. Over them, we seem to be dealing with the beginning of Würm I. First, there is a flora that is rather cold, but with 21 percent tree pollens, among them some deciduous trees. There quite quickly follows a cool and damp, temperate flora, with 30 to 35 percent tree pollens, many of them deciduous. This flora compares very well with the one found in layer 4 in Pech II, beneath level 4A. We are probably dealing with the equivalent of layers 4C2 (Typical Mousterian) and 4B (Denticulate Mousterian), which fits with the archaeological data.

One of the isolated blocks of breccia did undoubtedly belong to the Denticulate Mousterian. It gave 153 artifacts, 82 tools, with a Levallois index of 6.7, a scraper index of 12.2, and a denticulate index of 40.2. The other block brings another riddle. It was unfortunately small and not rich: 62 artifacts, 32 tools, among them 4 handaxes. But the style of the tools differs from the Mousterian of Acheulean tradition of Pech I, and they show the patination and slightly worn aspect of the flints of the other breccias. This raises the question of the possibility, in Pech I, of a Würm I Mousterian of Acheulean tradition.

UPPER PALAEOLITHIC MAN AT PECH DE L'AZÉ

Two Upper Palaeolithic implements were found in Pech de l'Azé. The first, an end scraper, was picked up on the ground the first time we went into Pech II (Fig. 23). The second, a burin, was found on top of the first breccia, covered by a very thin stalagmitic crust. Alas, it has since been lost. It is difficult, from two tools, to give more than an impression, but they did "feel" Magdalenian. They represent the traces of brief passages, at Pech de l'Azé, of Upper Palaeolithic man.

4. *Combe-Grenal*

We shall now leave Pech de l'Azé for a while, and turn to Combe-Grenal. Here again we have the testimonies of Jouannet and the abbé Audierne, although they are unfortunately less detailed than those for Pech de l'Azé.

THE EARLIER EXCAVATIONS

On August 11, 1816, Jouannet wrote to the comte W. de Taillefer:

> I went to explore . . . at the entrance of a dale called Combe Grenant, a very wonderful cave. The opening faces the South. All the inside is filled with the bones of birds and other animals, all mixed with a marl which fills the anfractuosities of the cave. . . . Inside the cave, on the ground, I was very surprised to find two of those Celtic flint tools. . . . At some distance from the cave, I found several remains of the same nature. . . . (18)

And the year after, in the *Calendrier de la Dordogne,* Jouannet gave further details:

> To the left, coming into a gully known as Combe Grenant . . . opens, obliquely, a cave which may be 10 to 12 feet deep and 8 feet high at the entrance. . . . It was dug by nature in a rather soft limestone, but the inside and the back are made of calcareous clay mixed with rock rubble. This clay is larded all through with innumerable bones of all sizes and kinds: of large and small birds, of quadrupeds of all sizes, even of men, if I am not mistaken. Everywhere I dug in the cave, I found bones. . . . (18)

On March 2, 1828, the abbé Audierne wrote a letter to his friend de Mourcin, in which he told him of his visit to Combe-Grenal:

> Climbing up to the Grenal cave, one finds some flints, but in very small quantity, and all of them rather roughly shaped. At the entrance of the cave, a few are to be found. This cave faces the Southwest. . . . The entrance is 9 feet 6½ inches. One cannot avoid noticing the traces of an old wall, from which I pulled out one tooth and one arrowhead.* From this entrance to a second wall, inside, is a distance of 17 feet, 6½ inches. All this part is open to the sky, but there is no doubt this is due to the collapse of the rock, some parts of which can still be seen. The width of the cave is 27 feet. Getting deeper into it, one can see the sinuosities of rock filled with a mortar made of bones, flints, and small stones. The back is also filled with a similar mortar. The wall, built of squared stones—and which today divides the cave into two parts, or rather seems to close it, since the first part is open—is posterior to the mortar which fills the sinuosities of the cave. . . . The Combe-Grenal cave is about 56 feet deep, 10 feet high, and 28 feet wide. Inside and around it, we found only roughly worked flints. . . . (19) (Fig. 29)

So it seems that the cave was already open in the early nineteenth century. However, at the end of Würm II, as we shall see, it was probably filled up to the ceiling, or perhaps a very small space was left under the roof on the right side (looking in). From these descriptions, and from what we saw before our excavations, it is clear that part of the deposits had been dug out before any scientific exploration, probably during historic times, and that a dwelling of sorts, or maybe a sheep pen, occupied the right part of the cave itself. A first wall was built to hold the pressure of the soil, and another inside the cave itself to close the dwelling or pen (Fig. 30). This last wall was still standing when we began our excavations.

In 1864, Lartet and Christy, writing a paper on the caves of Périgord, mentioned Combe-Grenal:

> In October 1863, we had some excavation work done in the back of this cave, which was then used . . . to store lime and some hay. It must, indeed, have been partially emptied at a more or less remote period, and what remains of the bone-bearing deposits did not permit us to get a definitive opinion as to how the mixture of bones and worked flints, haphazardly piled up in the explored part, came about. There we found the bones of hyenas (*H. spelaea*), wolves, foxes,

* Probably a pointed flake, or perhaps a Mousterian point.

Fig. 29

Combe-Grenal, six years after the excavations.

The view is taken from the lowest layers. On the left, part of the wall of the Acheulean cave is visible. The bottoms of the protective walls rest on the Würm I platform. Between the walls, one can see the slope toward the Würm II platform, and the present cave. The shelter extends on the left behind the unexcavated part.

> hares, horses, wild boars, red deer (*C. elaphus*), oxen, mountain goat, and chamois all mixed with worked flints . . . these, usually bad workmanship, were almost all of medium size and unvaried in shape. . . . It should be noted that this is the only cave we have explored in the Dordogne where we did not find any reindeer remains . . . . (23)

From this description we can assume that they explored the back of the cave, and probably found layers 10 to 16, which belong to the Denticulate Mousterian. It is, however a little odd that they did not find any reindeer remains, since this animal is present all through Würm II at Combe-Grenal.

Capitan also came to the Combe-Grenal at an unspecified time and picked up a small Acheulean handaxe, which is now in the Musée des Antiquités Nationales at Saint-Germain-en-Laye, near Paris. We have no information on the circumstances of its discovery, but it was probably picked up on the slope, for the Acheulean layers remained completely unknown until our excavations. In the 1930s Denis and Elie Peyrony did a small-scale excavation in the median layers (at a place where the upper ones had been destroyed long before) and picked up a rich series of tools. From the general aspect of the assemblages, now in the museum at Les Eyzies, they distinguished three layers, corresponding very probably to our layers 15 to 31.

In 1952, Denis Peyrony asked me to undertake new excavations to complete the stratigraphy, since as he put it, "I have been down to the bedrock but, paradoxically, have not excavated the top" (which, as we have seen, had been destroyed at this part before his excavations).

OUR EXCAVATIONS AT COMBE-GRENAL

So, in September 1953, we left Pech de l'Azé, thinking we could not do much more under the present conditions, and began the Combe-Grenal excavations. Since Peyrony, some pot-hunters had been at work, and the site, covered with brambles, did seem in rather bad shape. Peyrony had dug an L-shaped trench on the right side of the cave (looking in), turning inside to the left into the cavity dug out in "medieval" times (Fig. 30). The pot-hunters had enlarged it and dug horizontal holes into the richest layers.

We began by establishing a grid of square meters and initiated the excavation on the left side of Peyrony's trench, to establish a stratigraphy on a finer scale. It soon became apparent that the stratigraphy was indeed complex, if not particularly difficult. The site was layered like a cake, with different colors and textures (Fig. 31).

The following year, we dug on a wider surface and discovered that if Peyrony, at a point of his trench (X on the map), had really touched the bottom, at other places he had mistaken a slightly concretioned layer for the bedrock, and that there were other layers below. We penetrated deeper and arrived at layer 38, resting on bedrock. We already had seven more layers than Peyrony. The following years we searched for, and found, part

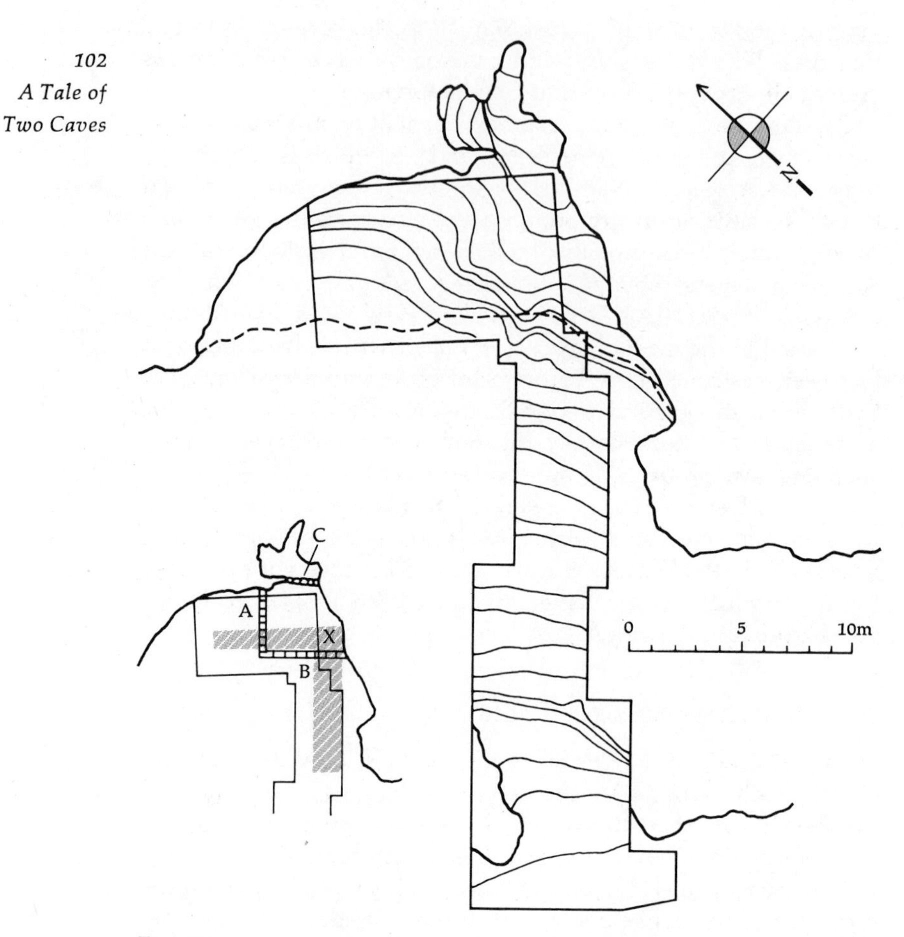

Fig. 30

Plan of Combe-Grenal showing the part excavated to the bedrock and the entrance to the Acheulean cave. On the left, the small-scale plan shows the part excavated by D. and E. Peyrony (shaded area) and the position of the walls (A, B, C) built when part of the sediments were destroyed. Wall C was still standing when we began excavating, but behind it the cave was practically empty. The dotted line represents the present roof of the shelter and X the place where the Peyronys touched the bedrock.

of these lower layers inside the cave, but there the series finished with layer 35, and layers 36, 37, and 38 did not penetrate into the present shelter. Layer 36 was reddish, clay-like, and probably the result of the weathering of the top of 37, which was made of brownish earth. At the same time, we had begun the excavation of the upper layers (1 to 14).

We thought at that time that the excavations were close to completion. But, in 1959, I had three American volunteers with me in the summer, and having no work to give them inside the shelter, I decided to enlarge the excavation toward the slope. A test dig, on a square meter, in E4, had already shown that below layer 38 there were probably other deposits, but in that part they were almost sterile. A series of big fallen blocks had prevented an extension of the test trench, but with this new enlargement, it would be possible to remove the blocks and go deeper. However, it did not look like a very promising enterprise.

How wrong we were!

Next year, we were able to push all the way down to layer 54 and added the whole Würm I sequence to the already well-known Würm II sequence. Under the poor and often cryoturbated layers of the top, we found three levels of ashes, often thick, filled with bones and flints. In 1961 we found a burial pit, empty, alas; and in 1962, having followed the bedrock to square V4, we thought that, this time, we had the whole sequence at Combe-Grenal. But on the last day of the excavation for the year—which was also to be the last day of excavations at Combe-Grenal—we found a small gully in the bedrock, filled with red dirt, and in this dirt some flakes and the tip of an Acheulean-like handaxe! We then remembered the handaxe found by Capitan and supposed that the bedrock had another step, analogous to the one separating the Würm I from Würm II. In 1963, we found out we were right. There was indeed another step. Under a thick reddish soil, corresponding to the last interglacial soil, were layers of yellow eboulis, with several Acheulean levels, dating back to Riss III. In 1964 and 1965, we followed the bedrock down to the point where the layers had been cut off by the erosion, and the bedrock was bare on the slope. Very probably, even older deposits existed once even further in front, but were destroyed by the erosion of the side of the small valley. For the lowest layers we found, we had only the extreme rear portions (Fig. 31).

THE CLIMATIC SEQUENCE AT COMBE-GRENAL

Thus, we now had a sequence of 64 layers, or rather 65, since layer 50 is divided into 50 proper and 50A. This sequence spans Riss III, and Würms I and II; and it completes and supplements the Pech de l'Azé sequence extremely well. The total thickness is close to 13 meters, about 43 feet.

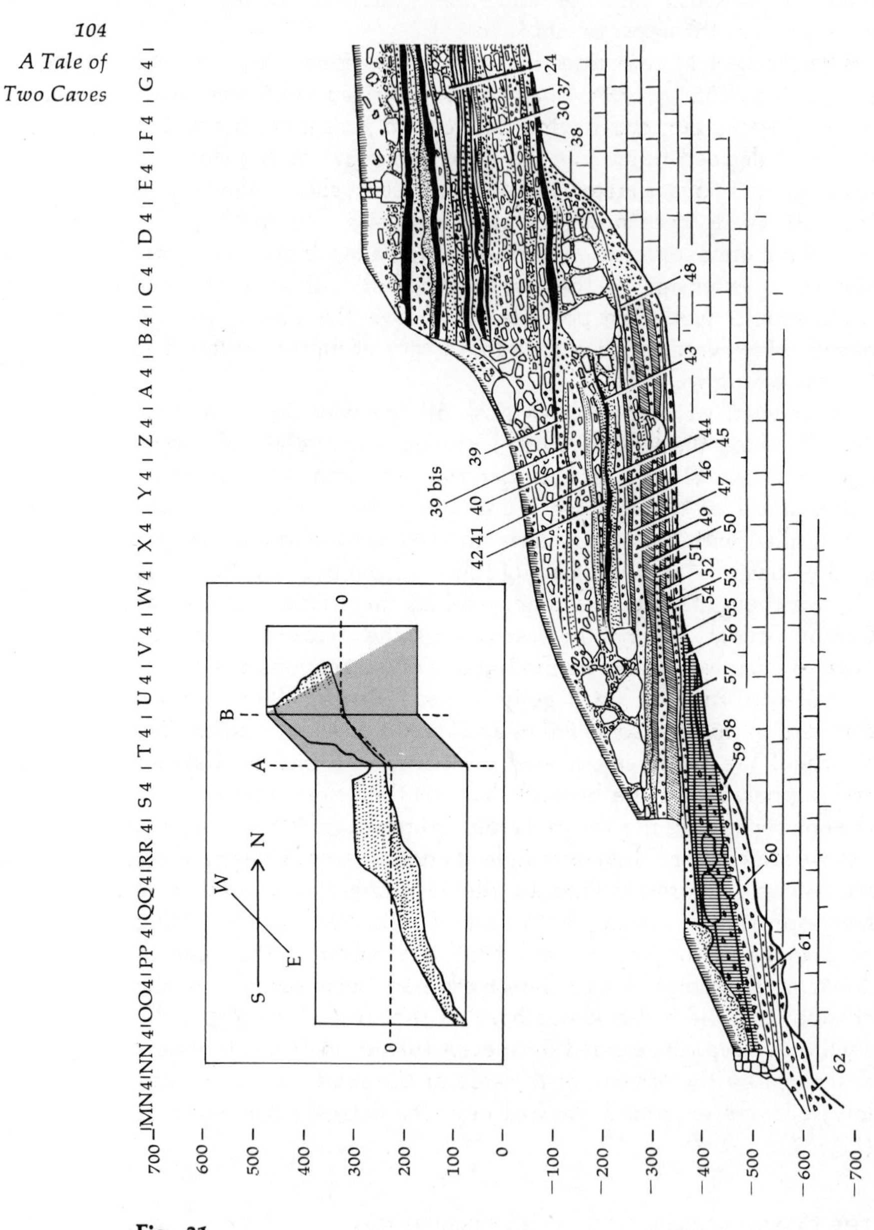

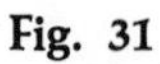

Fig. 31

Section of Combe-Grenal (semischematic).

The vertical shading in the lower half of the section represents the last interglacial soil, developed over the Acheulean layers. Layer 36 is the last Würm I layer.

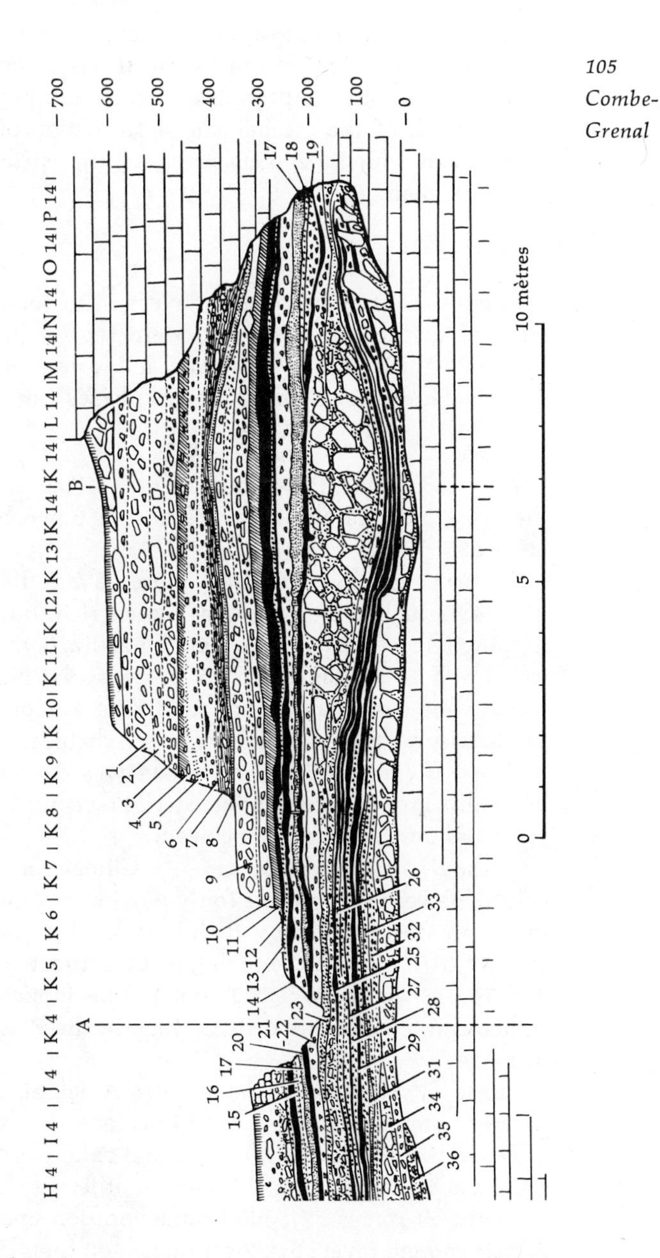

It is not possible, of course, to examine the sequence in all its details here, and in any case this is not the purpose of this book. But it can be broken down into 22 climatic phases, determined by sedimentary and pollen analysis, and into different archaeological phases, the same industry being often found in several consecutive layers.

Three major units are apparent, separated by steps in the bedrock: Riss III, Würm I, and Würm II. As a general rule, each is situated closer to the present cliff than the preceding one. This is the result of the mechanism of formation of rock shelters, a mechanism which was also active at the entrance of the caves (see Figs. 8 and 9).

Riss III

This consists of archaeological layers 64 to 56. The lowest layers, situated at the extreme front, were represented only by small remains, stuck to the bedrock, and often protected by their situation in hollows. In all, seven phases can be distinguished:

Phase I (layers 64 and 63): Very rich in angular congelifracts, which were deposited under slightly damp conditions at first, then dry and cold ones. Few trees: 5 or 6 percent tree pollens in layer 63.

Archaeologically speaking, these layers have yielded very little. Originally, they may have been rich, but we have almost nothing left of them. There were 64 artifacts in layer 64, 5 being tools, and 87 in layer 63, 15 being tools. No handaxes were found, but some of the special flakes are present, so we can safely assume that the industry was Acheulean.

Phase II (layer 62): The climate was damper, and probably less cold. Tree pollens 15 percent; 311 artifacts, among them 33 tools and one handaxe; Acheulean.

Phase III (layers 61, 60, and 59): Climate increasingly dry and cold; 7 percent tree pollens (only pine) and some steppe elements in the grasses and shrubs. Rich layers, which have given respectively 1,010 artifacts (56 tools), 6,800 artifacts (about 500 tools), and 16,623 artifacts (1,073 tools). The industry is a southern Acheulean, much more evolved than in the Riss I and II of Pech de l'Azé.

Phase IV (bottom of layer 58): A slight amelioration and greater humidity. Tree pollens 15 percent.

Phase V (top of layer 58): Dry and cold. Layer 58 is very rich: 17,432 artifacts, 1,409 tools. Same southern Acheulean as before.

Phase VI (layer 57): Slight amelioration and more humidity. A rich enough layer: 5,737 artifacts, 400 tools. Same industry as before.

Phase VII (layer 56): Cold and dry, steppic, with 7 percent tree pollens. Less rich: 1,113 artifacts, only 73 tools. Acheulean.

In all the Rissian layers at Combe-Grenal, easily the most predominant animal is the reindeer, with some wild goat, saiga antelope, red deer, and surprisingly enough, from time to time the fallow deer (*Cervus dama*), which is usually supposed to belong to a rather temperate climate. Some levels are also very rich in rabbit bones.

From the top of layer 59 and upward, we find the last interglacial soil, developed under the warm and damp conditions of the Riss-Würm, on the top of the Riss III deposits. But of course the implements, bones, and pollens found inside it date back to the time of the deposition of the sediments, that is, the final part of Riss III. As usual, we could not observe any interglacial deposits, with one possible exception. One of the first samples taken inside the soil for pollen analysis gave a surprising result, with wild vine pollens and other warm plants. But no other sample taken at the same level gave these results again. The pollens, nevertheless, seemed fossilized, not recent. And as there were some rodent burrows in the soil, it may well be that this unique sample represented the filling, during interglacial times, of one of these burrows. But in general the interglacials were periods of weathering and erosion, with few deposits, except in lakes, peat bogs, and tuffs.

As I said, the Acheulean at Combe-Grenal belongs to the southern variant. The flake component of the industry varies only slightly from layer to layer and looks like Mousterian of Acheulean tradition, a little roughly made, although from time to time some very beautifully made implements appear (Fig. 32). The method of flaking is not Levallois, and the maximum Levallois index, in layer 56, is only 8.4. The scraper index, in essential count, varies from 42 to 54, with very few Quina-like scrapers (QI maximum in layer 57 with only 1.3). Backed knives and Upper Palaeolithic types of tools are present, sometimes very typical, but not in particularly large proportions. Denticulates vary from 15 to 18 percent. The handaxe index varies from 5 to 10, of the same order as in the Mousterian of Acheulean tradition. The handaxe types are varied, but the classical types of the Upper Acheulean (lanceolate, elongated cordiform, Micoquian) are either rare or absent altogether. Many are badly made nuclei-

→

Fig. 32

Combe-Grenal. Acheulean (layers 58 and 60).

1, 2, Mousterian points. 3, 4, Denticulates. 5, Convex side scraper. 6, Naturally backed handaxe. 7, Amygdaloid handaxe. 8, Backed handaxe. 9, Backed knife.

1

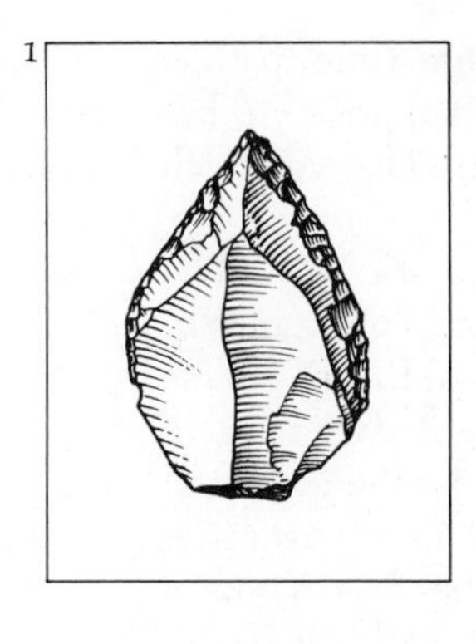

2

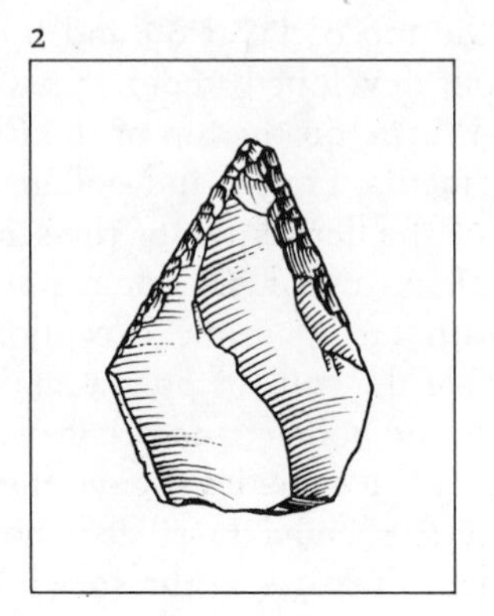

3

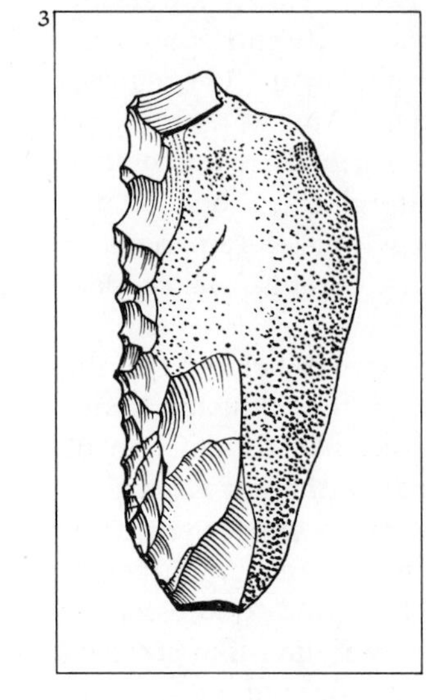

4

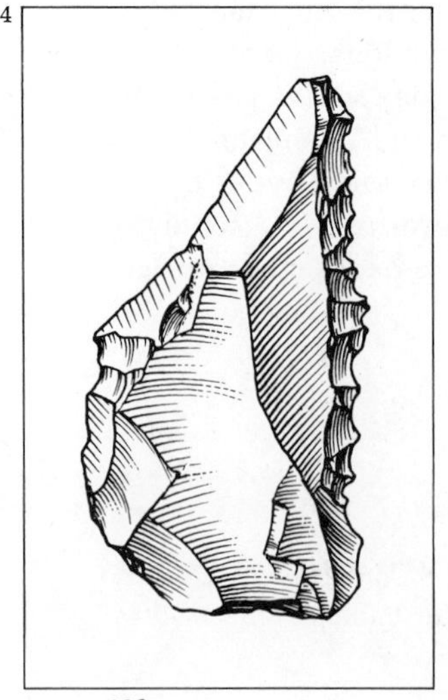

5

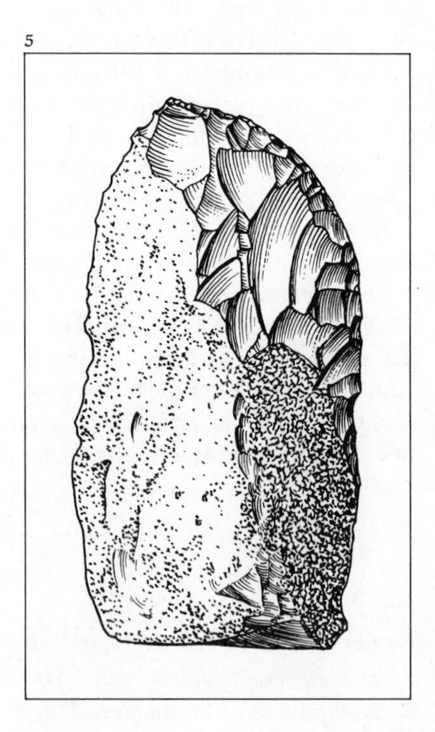

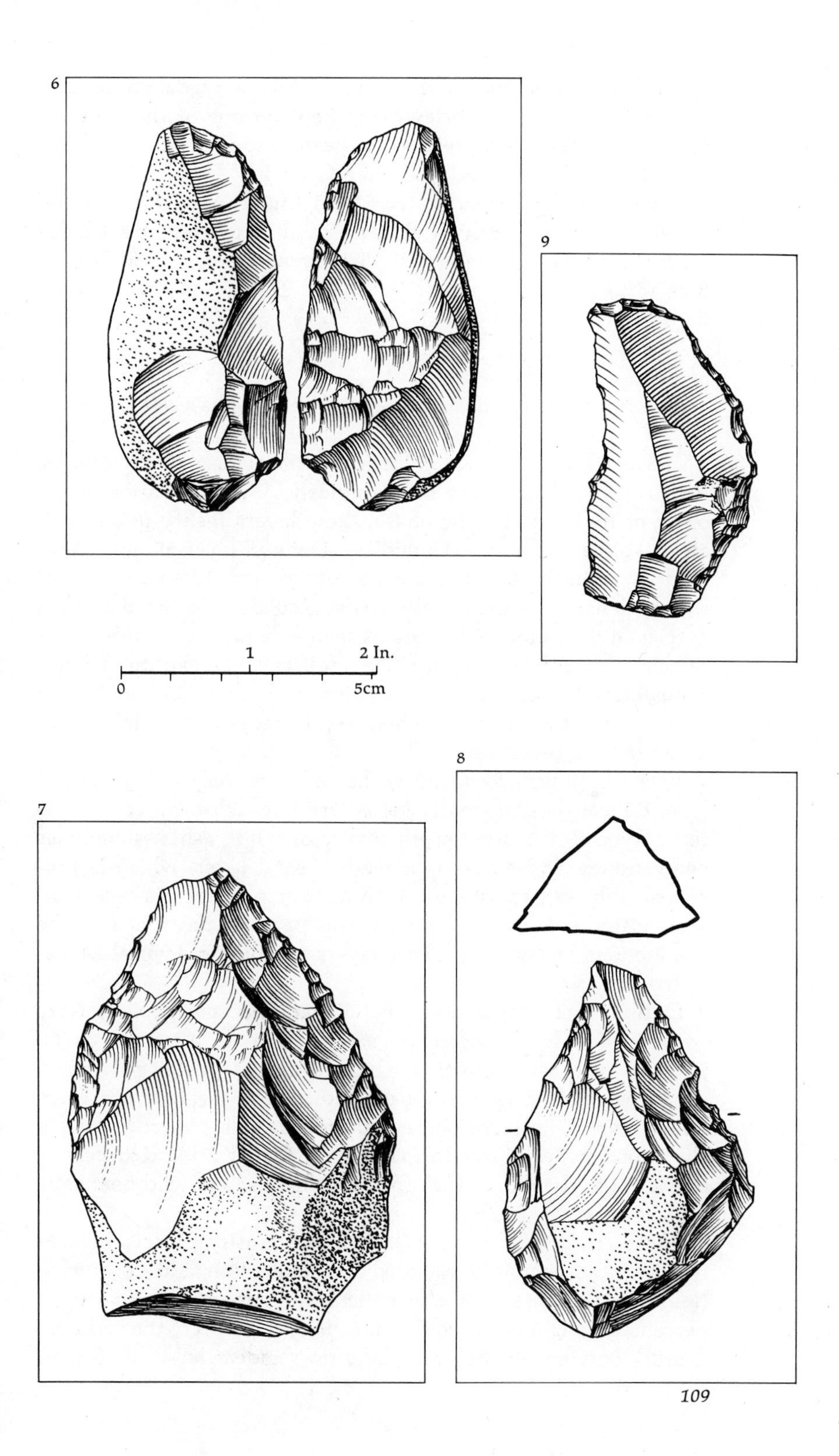
6
9
1
2 In.
0
5cm
7
8

form. When well made, they are mainly amygdaloid or also backed handaxes, the cortex being kept on one of the edges or this edge being truncated by a steep retouch. No true flake cleavers have been found at Combe-Grenal.

A very similar Acheulean has been found in an open-air site about 100 miles south of Bordeaux, at Bouheben, by Claude Thibault of the Laboratory of Pleistocene Geology and Prehistory at Bordeaux.

Würm I

This can also be divided into seven climatic phases (Fig. 13):

Phase I (layers 55 to 53). Rich in thermoclastic fragments, which have subsequently been rounded by cryoturbation at the onset of the next climatic phase, these layers testify to progressively colder conditions. Humidity, strong in layer 55, gives way to drier conditions. Trees are not numerous (12 to 15 percent tree pollens) and are mostly pines. Red deer is the dominant species in the fauna, with horse, roe deer, and so on; no reindeer remains have been found. The industries belong to Typical Mousterian but are very much crushed in layers 55 and 53 and only a little less so in 54, where some traces of fire and lenses of ashes have been found.

Phase II (layers 52 to 50A). Layer 52 is very rich in ashes; layer 51 consists of debris secondarily rounded by cryoturbation; layer 50 is made up principally of white ashes, sometimes concretioned; layer 50A is a reddish sand layer, which is preserved only in part of the site now, but probably (as we shall see) existed everywhere before it was partially destroyed by the solifluctions of layer 49. These layers all contain Typical Mousterian.

During their deposition, the climate became progressively warmer. The level of tree pollens goes up to a very high 60 percent, with a composition which is almost postglacial. However, the relative rarity of oaks indicates a mean temperature lower than the present one in this region. Also, there is only a slight pedogenesis (soil formation) on top of 50A. Red deer is dominant in the fauna, with roe deer, bovids, and wild boar. The reindeer is very rare, but present.

Phase III (layers 49 to 44). Consists mostly of accumulated fragments, secondarily rounded by solifluction and cryoturbation. The percentage of tree pollens falls down to 24, then 14 percent. The climate is colder and drier. However, the red deer is still dominant in the fauna, and no reindeer has been found.

Note the presence of Merck's rhinoceros in layer 49. The artifacts are very much battered, and it is impossible to tell which type of Mousterian was present, but there is nothing to indicate that it cannot be Typical Mousterian.

Phase IV (layers 43 to 41). The sediments here are different, with more sands and clays, reddish in color. The eboulis are not only blunted, but also porous (from weathering). The carbonates have been leached out, and there is a concentration of clay in the lower part. These characteristics indicate the beginning of soil formation. The climate is damp and temperate. The percentage of tree pollens goes up to a maximum of 60, with numerous deciduous trees. This oscillation is comparable to the one in phase II, but probably a little warmer. However between layers 41 and 42 there is a thin bed of angular fragments, almost without dirt between them, which could indicate a short but strong cold oscillation. Unfortunately, it was not possible to obtain pollens out of them. Red deer is dominant in the fauna. The industries are a kind of Typical Mousterian, very rich in Levallois flakes, different from the one in layers 50A to 55, for layers 42 to 43, perhaps a Denticulate Mousterian for layer 41, which is poor.

Phase V (layers 40 and 39). From the sedimentological and palynological data, it is clear we again have a colder and very much drier climate. Only 13 percent tree pollens, mostly pines. Shrubs and grasses of steppic character appear. The fauna consist of red deer, horse, and reindeer. In layer 40, we have Typical Mousterian, not very rich. In layer 39, the implements have been battered subsequently by cryoturbation, and it is difficult to tell for sure which kind of Mousterian they belong to, but probably Typical Mousterian here also.

Phase VI (layer 38). The conditions are less cold and the percentage of tree pollens goes up again, but only to 33, with the presence of some deciduous types. The steppic elements disappear. The climate is damp, and this is probably the reason for the cryoturbation of layer 39. In the fauna, the red deer is dominant. The industry is a Denticulate Mousterian, rich in Levallois flakes.

Phase VII (layers 37 and 36). Very thermoclastic, these layers correspond to a cold climate; tree pollens are only 11 and 10 percent, with no deciduous trees. The steppe elements are more numerous than in layers 39 and 40. The reindeer is present in the fauna, and the industry is Typical Mousterian.

On top of these layers is a soil, developed during the interstadial which followed.

Würm II

There are eight climatic zones in Würm II, but with less marked oscillations than during Würm I:

Phase I (layers 35 to 26). Climate extremely cold and very damp. Only 7 to 5 percent tree pollens, most of the time pine only. Red deer is still dominant in layers 35 and 34, then bovids in 33 and 32, then reindeer. The types of industries vary: Ferrassie (35 to 32), Typical Mousterian (31 to 28), Ferrassie again (27), then Quina (26).

Phase II (layers 25 to 23). Climate very cold and dry. Less tree pollens than in the preceding layers; only some pines are left. Reindeer is dominant. Quina Mousterian.

Phase III (layers 22 to 20). A slight amelioration, and augmentation of humidity. The percentage of tree pollens goes up to 16; besides pines, there are some deciduous trees, such as hazel and alder. Reindeer is dominant. Quina-type Mousterian in layers 22 and 21, Denticulate Mousterian (non-Levallois) in layer 20.

Phase IV (layers 19 to 14). A very low percentage of tree pollens, numerous steppic elements, climate very cold and dry. At the beginning, the sediments are loaded with eolian elements. Reindeer dominates in layers 19 to 17, then horse in layers 15 and 14. This change could be more cultural than ecological, since layers 19 to 17 are Quina Mousterian, and layers 16 to 14, Denticulate Mousterian.

Phase V (layers 13 to 11). The climate is damper, with probably a slight thermic amelioration. Tree pollens form up to 17 percent. Horses and bovids dominate. Denticulate Mousterian.

Phase VI (layers 10 and 9). The percentage of tree pollens falls below 10. The climate is cold and dry, but less steppic than in phases II and IV. Bovids and red deer dominate over reindeer in the fauna. The assemblages from these layers are pretty poor, but seem to belong to Typical Mousterian, mainly layer 10.

Phase VII (layers 8 and 7). About 15 to 18 percent tree pollens with, besides pine, some deciduous trees. The climate is less cold, but damp, as indicated by the leaching of carbonates. Red deer and reindeer dominant. Layer 8 is very poor (44 tools, 356 artifacts), but could belong to Denticulate Mousterian (in essential count, scrapers formed 9 percent and denticulates 36 percent), but from so few tools it would be unwise to be too affirmative. Layer 7, comparatively rich (802 artifacts, 193 implements), is a small-sized but well-made Typical Mousterian with plenty of Levallois flakes.

Phase VIII (layers 6 to 1). The percentage of tree pollens again goes below 10, with numerous steppic elements. Climate very cold and very dry. Layers 6 and 5 are poor, probably belonging to the same type of Typical Mousterian as layer 7; but with layer 4, we have a change in the industry. This layer is also poor (138 artifacts, 21 tools). At that time the shelter was almost completely filled and could be used only for short stays by a few men, maybe during hunting. But among the tools there is a typical backed knife, and among the flakes are at least four which are unmistakably bifacial work flakes. So it seems that the Mousterian of Acheulean tradition begins with this layer. Layer 3 has two backed knives and two handaxes out of a total of 16 tools. Layer 2 has one backed knife and some typical flakes, and layer 1 has six backed knives and two handaxes. At that time there was practically no shelter left open, and one certainly could not stand up under the roof.

From this long sequence of climatic and archaeological variations, we can readily see that if there is any correlation between climate and tool kit, it is a very loose one indeed. We shall have more to say about this when we compare Combe-Grenal with Pech de l'Azé.

SOME COMMENTS ON THE MOUSTERIAN INDUSTRIES AT COMBE-GRENAL

This is not the place to analyze in detail all the different Mousterian assemblages throughout Würms I and II. However, some comments may be useful. Most of the assemblages fall easily into the divisions we have set for the Mousterian, and when there is difficulty in assigning a layer to a given type of Mousterian, it is either because the layer is too poor, and statistical fluctuations could play too great a role, or because the implements have been battered beyond recognition.

Within each type of Mousterian there are interesting variations.

Typical Mousterian

The first layer that can be attributed to Typical Mousterian is situated at the very bottom of the Mousterian sequence, either directly on the bedrock or on the interglacial soil. It is mixed with rounded eboulis, rounded by cryoturbation and not dissolution, since they are not porous. Many of the flints and bones are crushed, too. This layer has yielded only 48 recognizable tools, 37 in essential count. This is one of the cases when one should be careful; but by comparison with the rich layers above, either uncryoturbated or less cryoturbated, it can be assumed that layer 55 belongs to the Typical Mousterian.

There is a unity in these layers (55 to 47). They have a Levallois index of medium value, and the faceting indexes are also medium. The scraper index is high without being very high, the Quina index is high for Typical Mousterian, and the denticulates are generally well represented. Mousterian points are usually present in a fair percentage. These layers compose that we call the lower complex of the Typical Mousterian at Combe-Grenal (Fig. 33).

Layers 42 and 43 (43 is a hearth level at the bottom of 42) have quite different characteristics. The scraper index is much lower (less than 30 in essential count), the Quina index is null, and the Mousterian points are sometimes absent. In layers 42 and 43, the typological Levallois index (in real count) is unusually high (59.6), even if the technical Levallois index is not much higher than in the other layers. This forms the second complex of Typical Mousterian (Fig. 34).

Layer 36, which ends the Würm I sequence, is different again, with a higher Levallois and laminary index. For the scraper index and the Quina index, it is closer to the lower complex than to the second complex.

Würm II begins with a Ferrassie series, which we shall examine later. From layer 31 to layer 28, we find a third complex of closely related industries (Fig. 35). Layer 31 is poor, but fits easily with the ones that follow. As a rule, this third complex has a high Levallois index, high faceting indexes, a good laminary index, and a high scraper index. The Quina index varies from 0 in layer 31 (but this may be a fluctuation) to 2.3. Mousterian points are well represented, denticulates less numerous than in the preceding Typical Mousterians. It should be noted that layer 36, at the end of Würm I, could easily fit into this complex, except for a higher proportion of denticulates.

Then, for a very long time, Typical Mousterian disappears from the sequence, and we have to get to layer 10 to find it again. This layer recalls the ones in the second complex of Würm I. Layer 7, on the other hand, is quite close to the third complex, but less "faceted." It has an unusually high typological Levallois index. Layer 6 would probably fit with 7 (the lower scraper index being the result of a statistical fluctuation in a comparatively small assemblage), for judging by the other characteristics and the style, it is very close to it (Fig. 36).

Thus, during Würm I and II, there are two main subtypes of Typical Mousterian at Combe-Grenal, one with a rather high scraper index, the other with a lower one. This reproduces the splitting of Typical Mousterian found in other sites and probably means that in the future a subdivision will have to be made.

Ferrassie Mousterian

The Ferrassie Mousterian is limited to the first part of Würm II, where it interstratifies with Typical Mousterian. Layer 35, which begins Würm II, has a high Levallois index, medium high facet-

on pages 116 and 117 ⟶

Fig. 33

Combe-Grenal, Typical Mousterian, Würm I (layer 50).

1, Levallois flake. 2, Levallois point. 3, Mousterian point. 4, Double side scraper. 5, Borer. 6, Convex side scraper. 7, Transversal scraper. 8, Convergent scraper. 9, Straight side scraper. 10, End scraper. 11, Side scraper with a thinned back. 12, Denticulate.

on pages 118 and 119 ⟶

Fig. 34

Combe-Grenal, Typical Mousterian (layer 42–43).

1, 2, 3, Levallois flakes. 4, Mousterian point. 5, Straight side scraper. 6, Convex side scraper. 7, Side scraper on bulbar face. 8, Notch. 9, End scraper. 10, Denticulate.

on pages 120 and 121 ⟶

Fig. 35

Combe-Grenal, Typical Mousterian (layer 29).

1, Levallois flake. 2, 3, Mousterian points. 4. Simple straight side scraper. 5, Convergent scraper. 6, Transveral scraper. 7, Double side scraper. 8, End scraper. 9, End denticulate.

on pages 122 and 123 ⟶

Fig. 36

Typical Mousterian, Combe-Grenal, Würm II (layer 7).

1, 2, Levallois flakes. 3, Mousterian point. 4, Double side scraper. 5, 6, Simple side scrapers (No. 5 has some accommodation retouches near the butt). 7, "Raclette." 8, Double concave scraper. 9, Denticulate. 10, Levallois core. 11, Déjeté *scraper.*

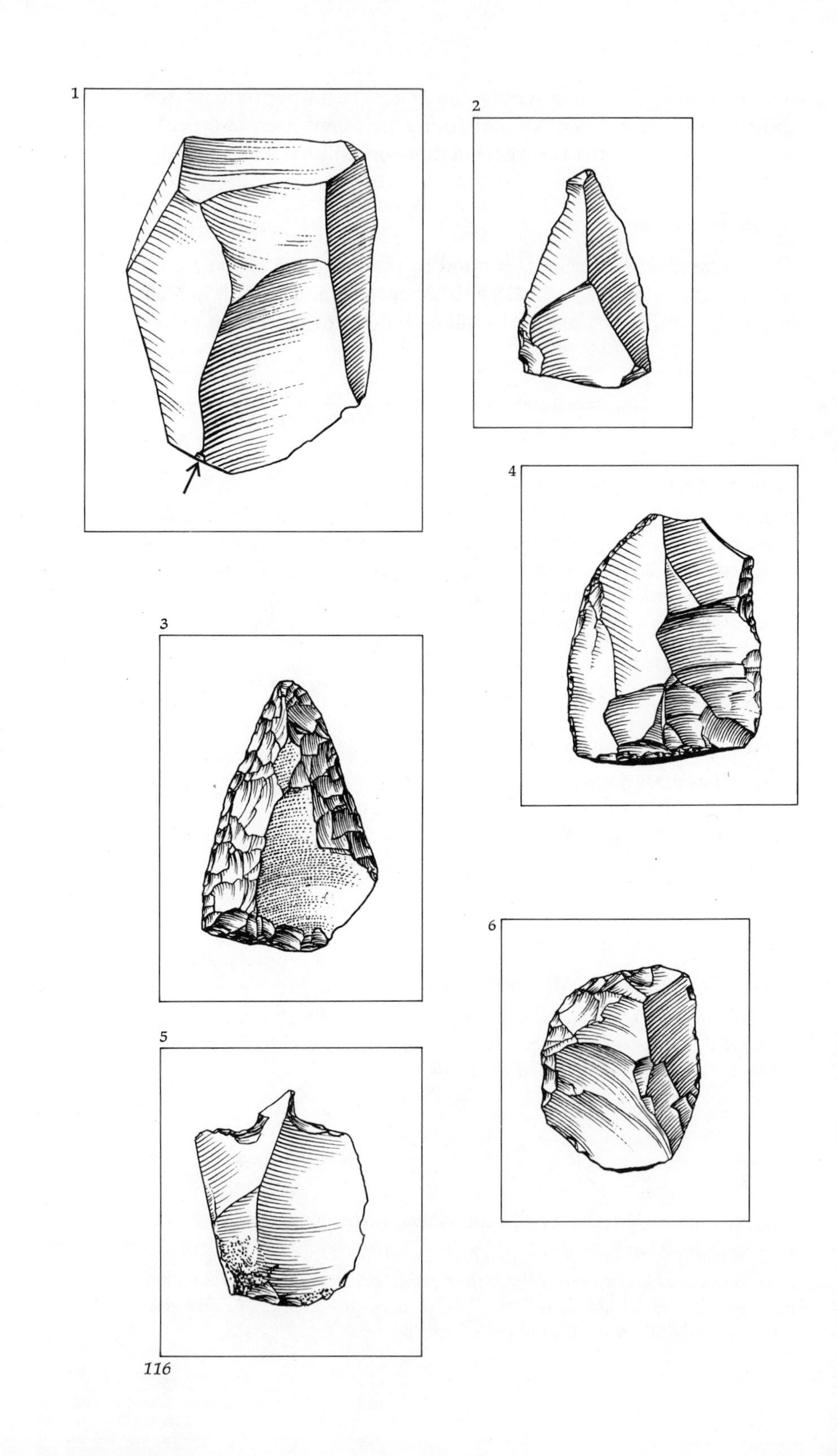
1
2
4
3
6
5

7

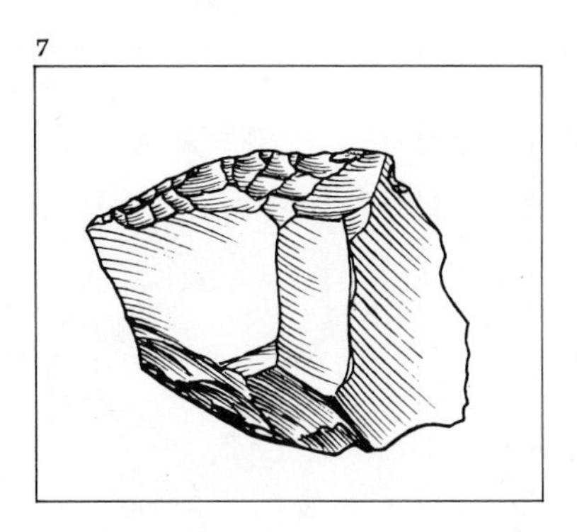

8

9

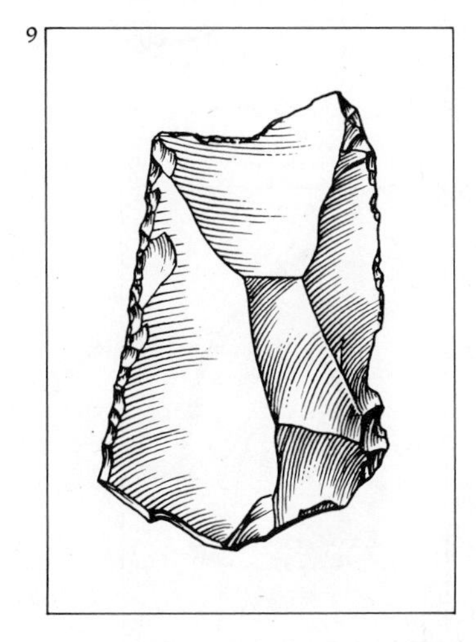

10

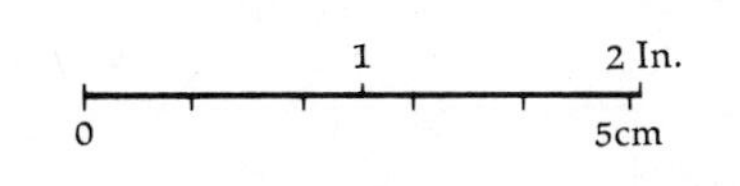

12

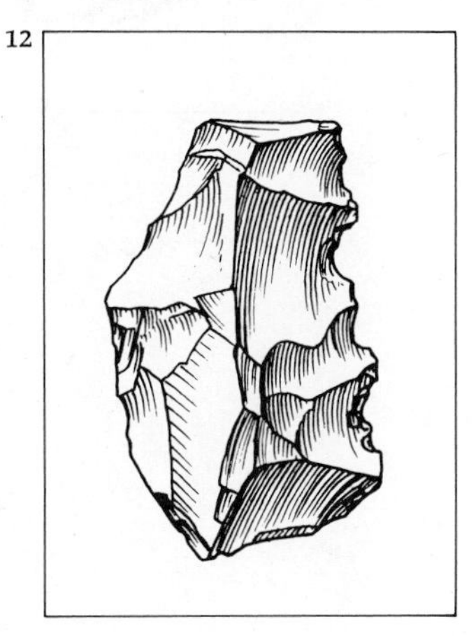

11

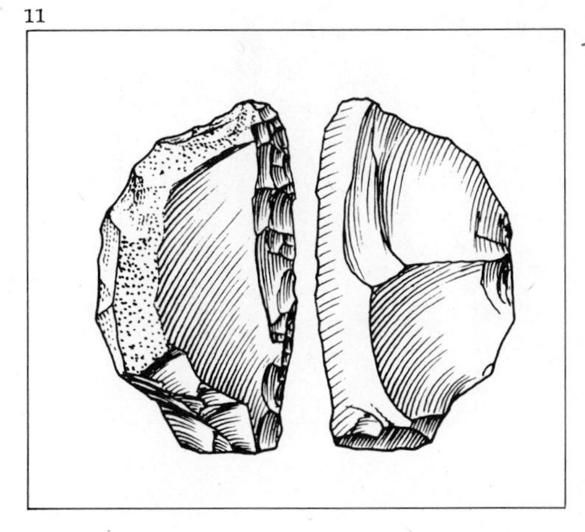

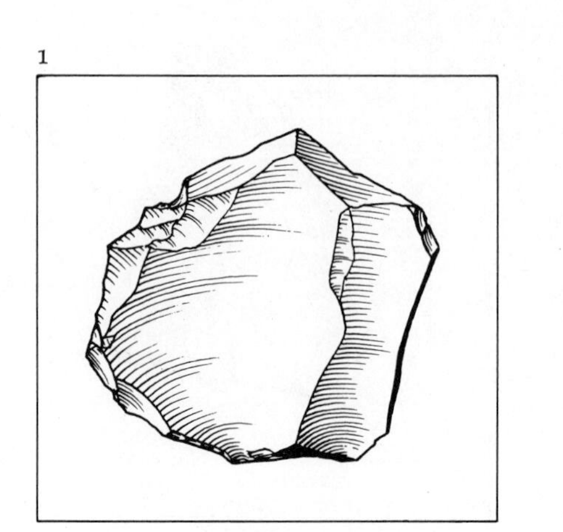
1

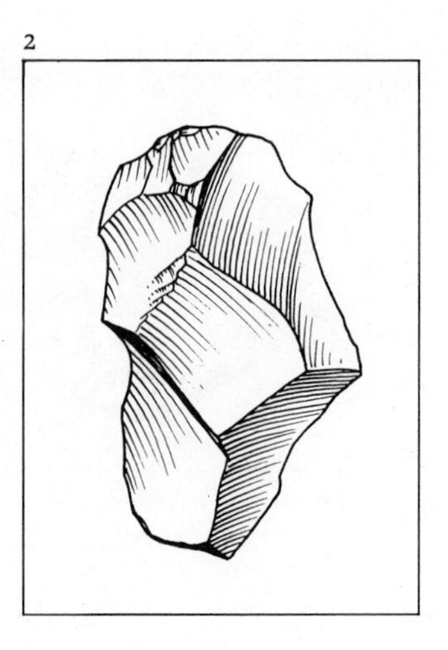
2

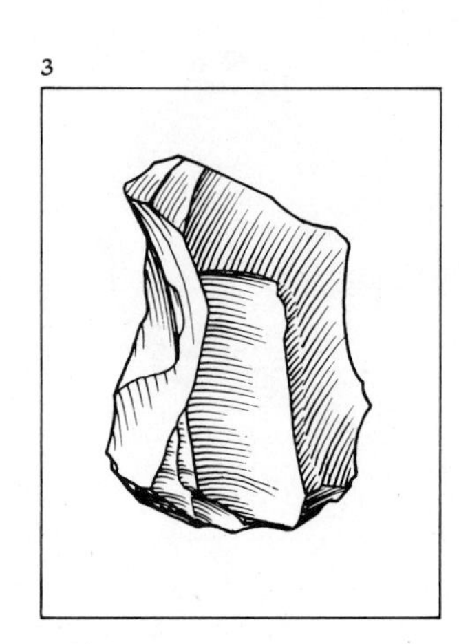
3

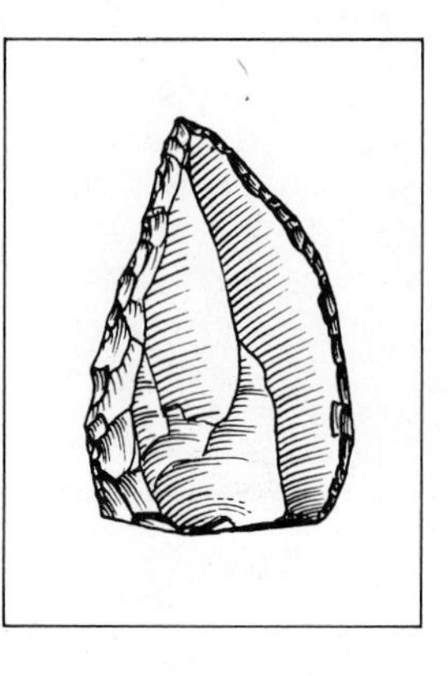
4

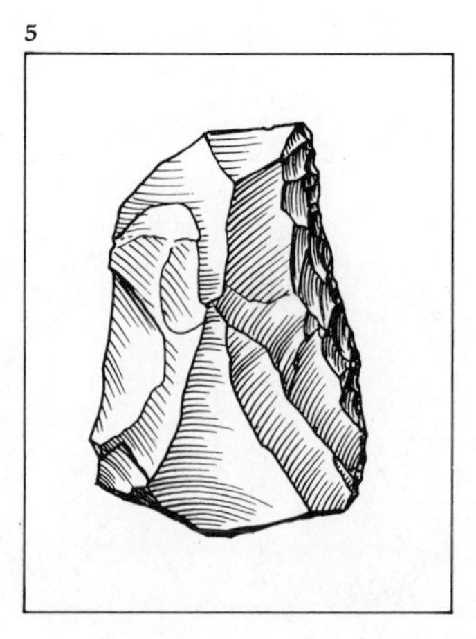
5

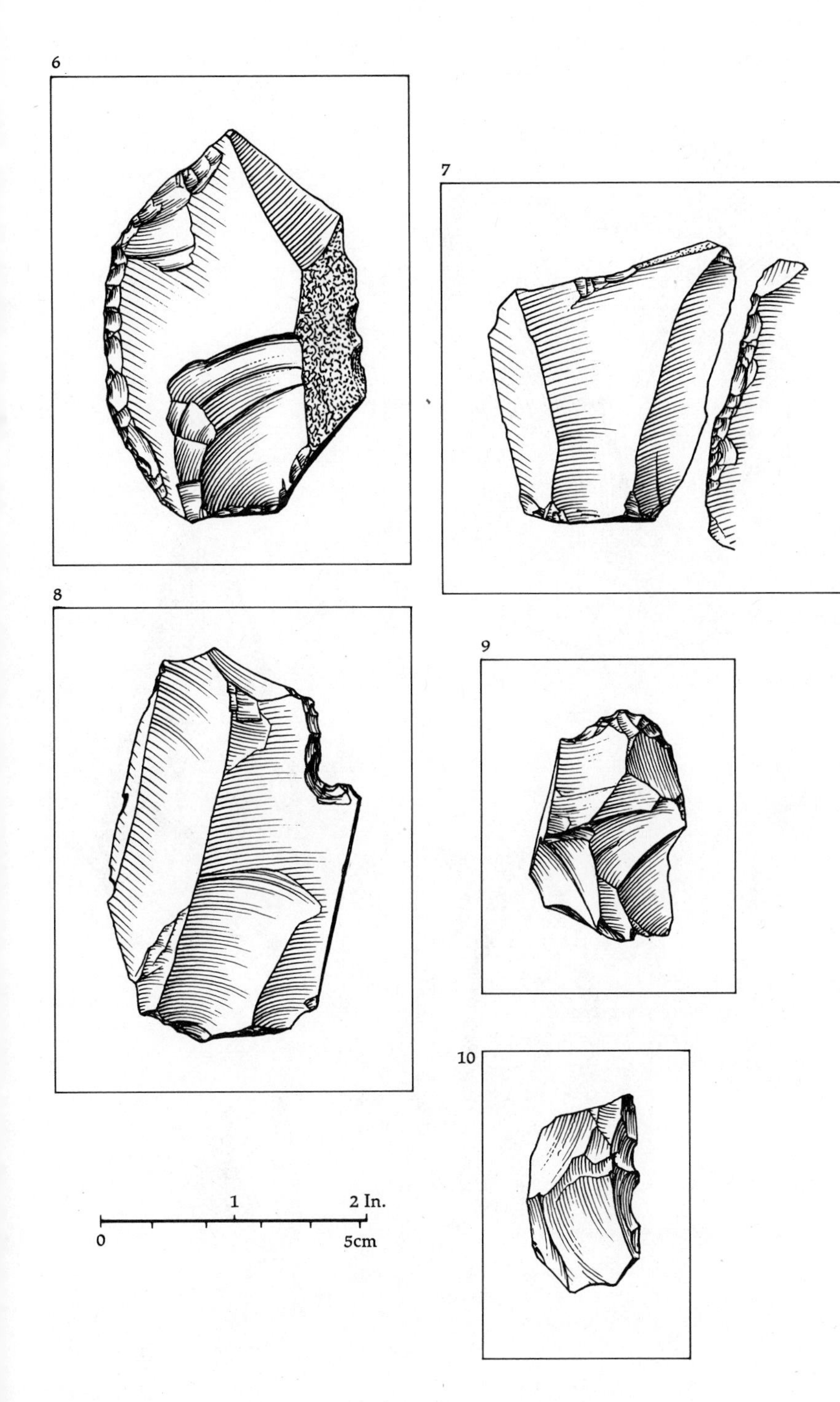
6
7
8
9
10
1
2 In.
0
5cm

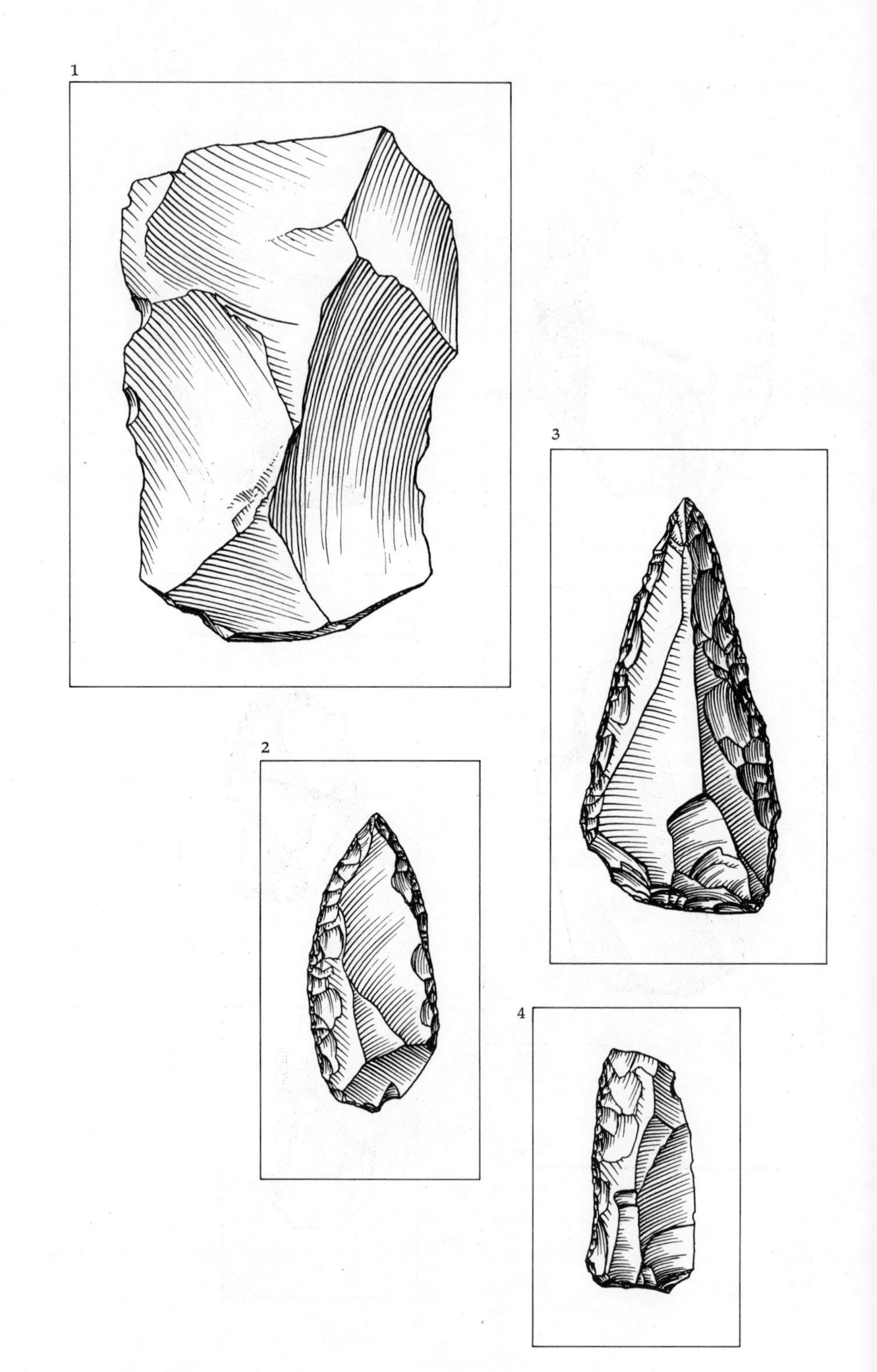
1
2
3
4

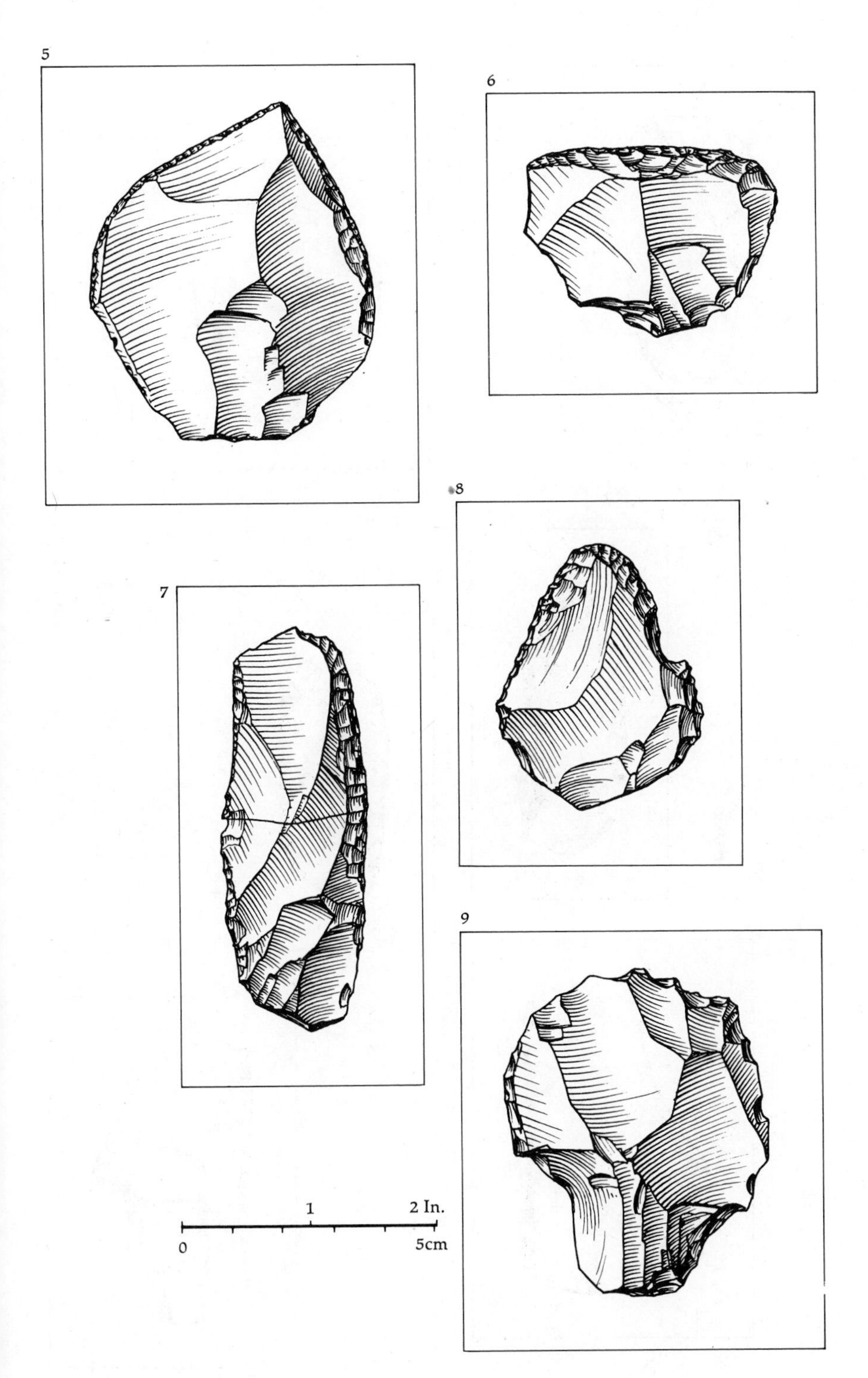
5
6
7
8
9
1
2 In.
0
5cm

1

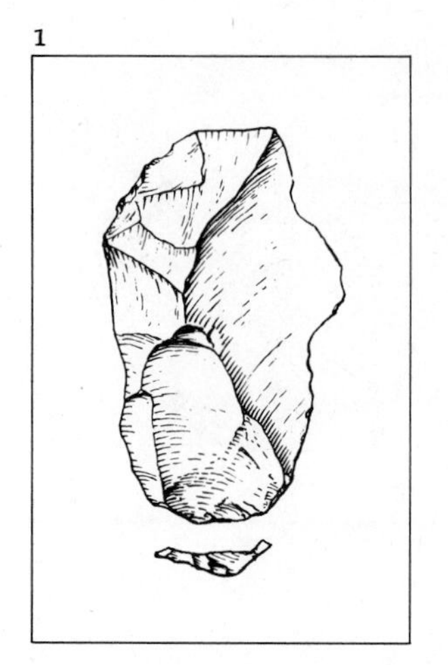

2

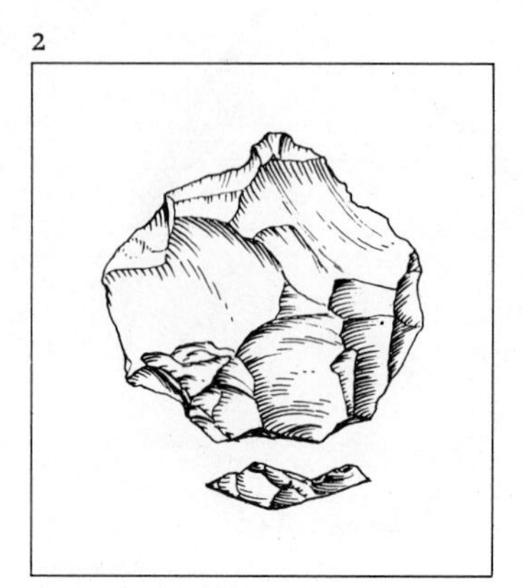

3

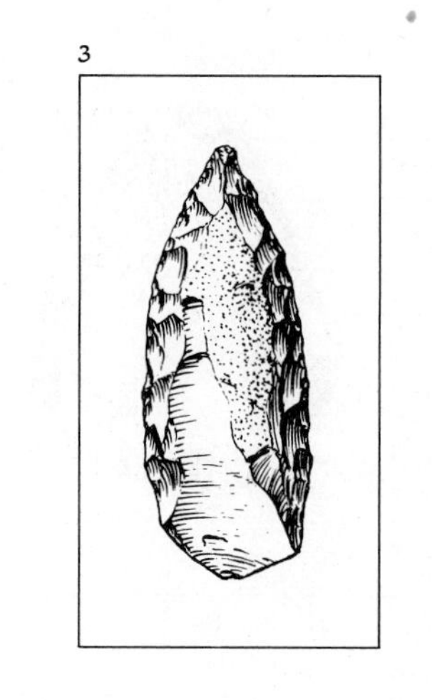

4

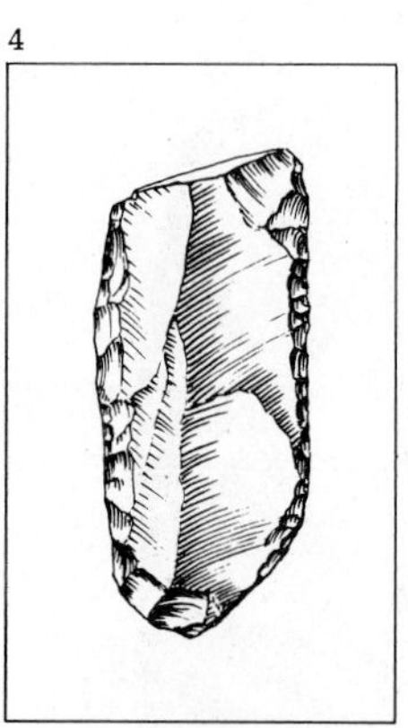

5

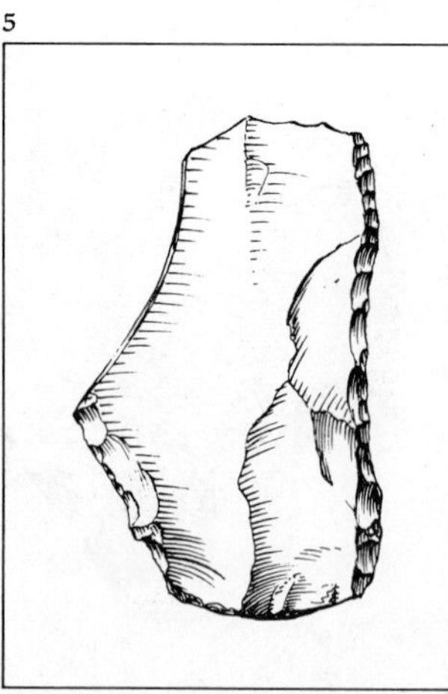

6

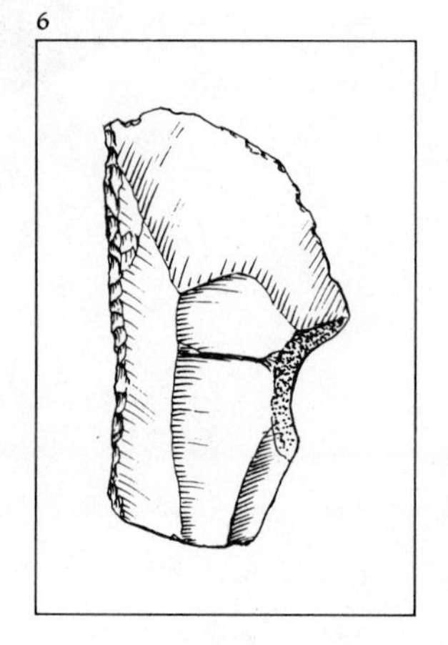

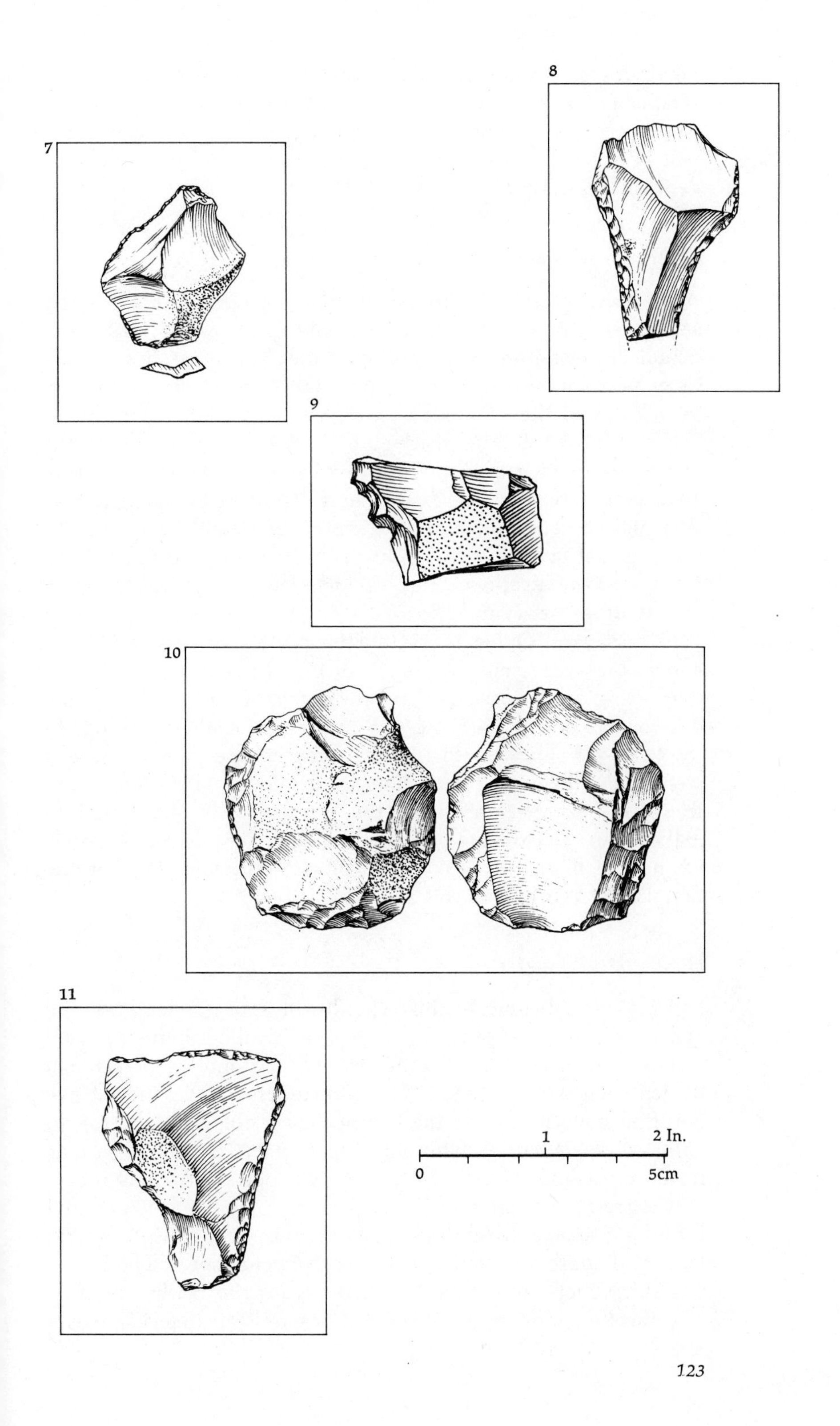
7
8
9
10
11
1
2 In.
0
5cm

ing indexes, a high laminary index, and a very high (78.3) scraper index. The Quina index is 14.3. Mousterian points are well represented. These characteristics, with some variations, will be found in the other Ferrassie layers. The percentage of denticulates is always low (Fig. 37).

Quina Mousterian

After layer 27, the Quina Mousterian develops for a lengthy sequence of nine layers, uninterrupted except for layer 20, Denticulate Mousterian. The Levallois index is always low, sometimes very low (layer 19, LI = 0.8), the faceting indexes low to very low, and the scraper index high to very high (up to 80.5). There is an evolution, marked by a general lowering of the scraper index (which in the last layer, 17, is only 50.2), and by an augmentation of the denticulates. The Quina index is also very high, except in layer 17, where it is of medium value (12.8). But this last layer is clearly Quina by its flaking technique and the very low Levallois index. Mousterian points are scarce or absent altogether (Fig. 38).

Ferrassie and Quina characteristics overlap at Combe-Grenal, and as they represent two facies of one industry—the Charentian—it is quite possible that one derives from the other, the difference being mainly the percentage of Levallois flaking. In the Ferrassie subtype, thick Quina scrapers are rarer because it would be difficult to make them on a thin Levallois flake; and transversal scrapers are rarer too, because it is impractical to make them on elongated Levallois flakes. Seen in these terms, we have a huge Ferrassie-Quina block occupying most of the first part of Würm II of this site.

Denticulate Mousterian

Layer 41 may belong to this type, but it appears clearly at the end of Würm I, with layer 38. It is a Levallois industry, well faceted, where the scraper index is low (7.6) and the percentage of denticulates high (33.6). No Mousterian points; indeed, we will find none in any of the Denticulate Mousterian layers; no Quina scrapers, either, with one exception. No true handaxes, as in the preceding industries. A high typological Levallois index.

Denticulate Mousterian then disappears from Combe-Grenal for a long time, and we have to get to layer 20, high into Würm II, to find another example, of rather different aspect. The Levallois index is at a minimum for all the industries of the site: 0.5. The faceting indexes are very low, as well as the typological

Levallois index, but the percentage of denticulates is very high (43.4). The scraper index is very low (9.5), and even if the flaking technique is more or less similar to that found in Quina levels, the Quina index is only 2. Further, we are not even sure that these Quina scrapers actually belong to layer 20, since this one is sandwiched between two Quina Layers (21 and 19) and they could very well represent a contamination.

Then we find a block of Denticulate Mousterian levels, from 16 to 11, without interruption, but in no ways quite homogenous. Layer 16 is again Levallois, faceted and rather puzzling: the scraper index is low (14.0) but the percentage of denticulates is not overly high (20.6); the Upper Palaeolithic types of tools are well developed. But layers 15, 14, and 13 form a unity, with a low to very low scraper index and a high to very high percentage of denticulates. Layers 12 and 11 see a new development of Levallois technique, particularly in 11 (Fig. 39).

on pages 126 and 127 ⟶

Fig. 37

Combe-Grenal, Ferrassie-type Mousterian (layer 27).

1, 3, Levallois flakes. 2, Convergent scraper. 4, Double side scraper. 5, Déjeté *scraper. 6, "Limace." 7, Mousterian point. 8, Denticulate. 9, Double concave side scraper. 10, Burin.*

on pages 128 and 129 ⟶

Fig. 38

Combe-Grenal, Quina Mousterian (layers 22 and 23).

1, Déjeté *side scraper. 2, Quina-type convex side scraper. 3, Denticulate. 4, Utilized bone: the tips are worn smooth; scratches on the side indicate utilization as a flaker. 5, Quina-type transversal scraper. 6, Quina-type bifacial scraper. 7, "Limace."*

on pages 130 and 131 ⟶

Fig. 39

Combe-Grenal, Denticulate Mousterian (layer 13).

1, Side scraper. 2, Natural backed knife. 3, Borer. 4, 6, 7, 8, Denticulates. 5, Atypical end scraper. 9, Notch.

Bottom: Cast of the posthole found in square F-8. Total length of the cast, 21 cm (8 ins).

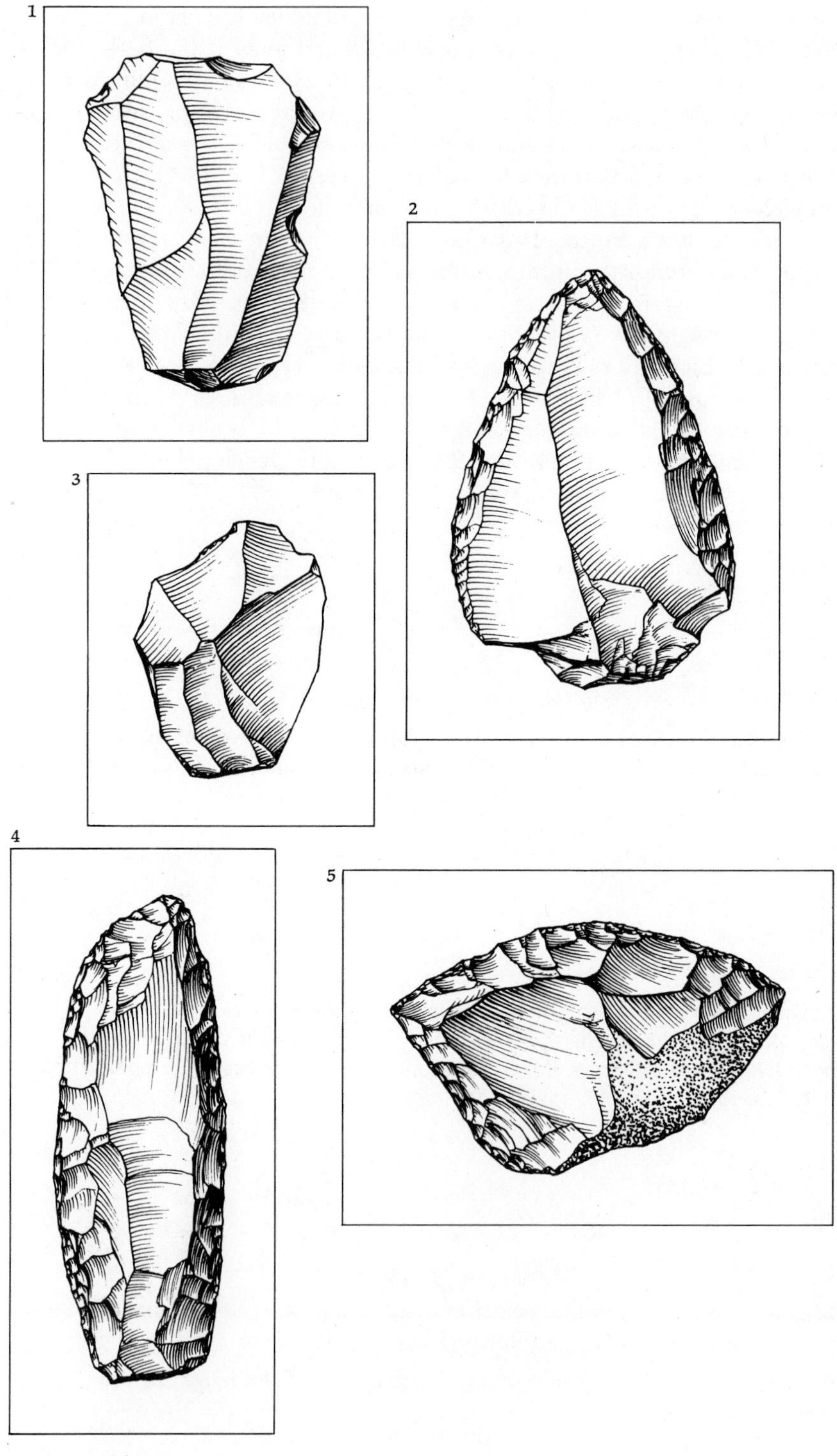
1
2
3
4
5

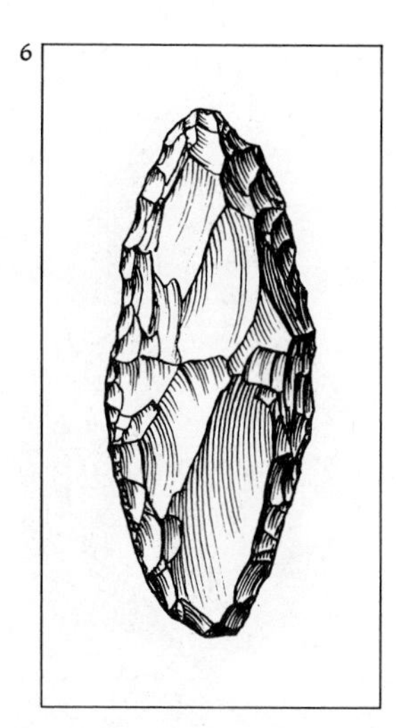
6

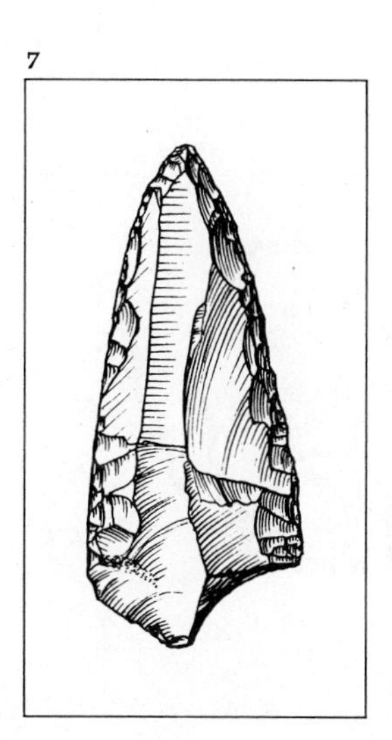
7

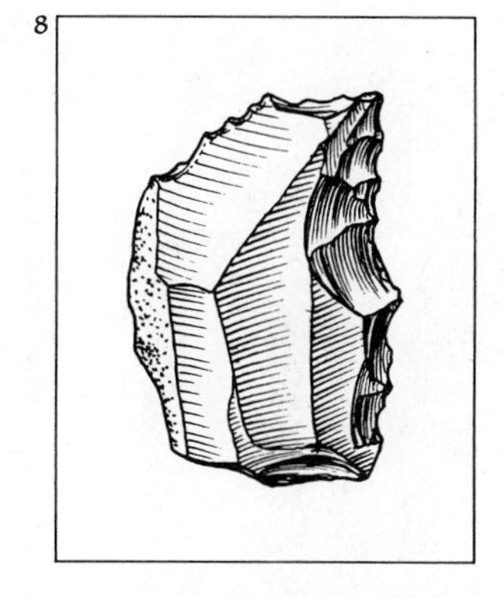
8

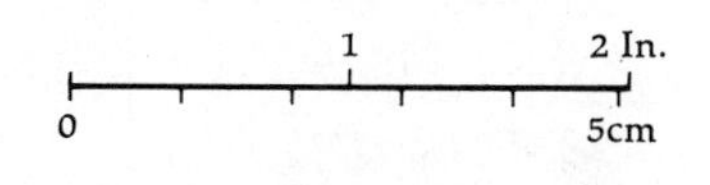
1
2 In.
0
5cm

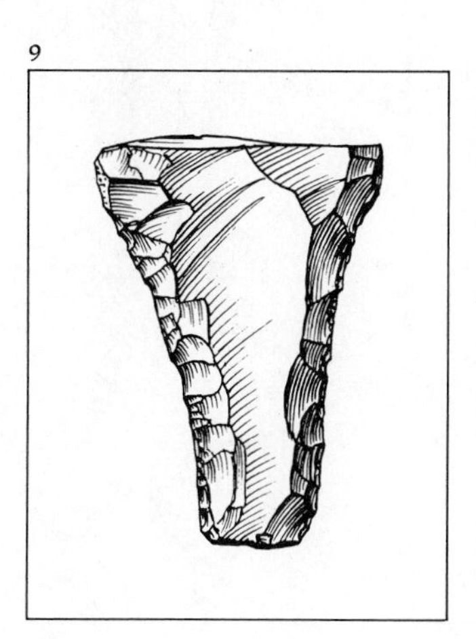
9

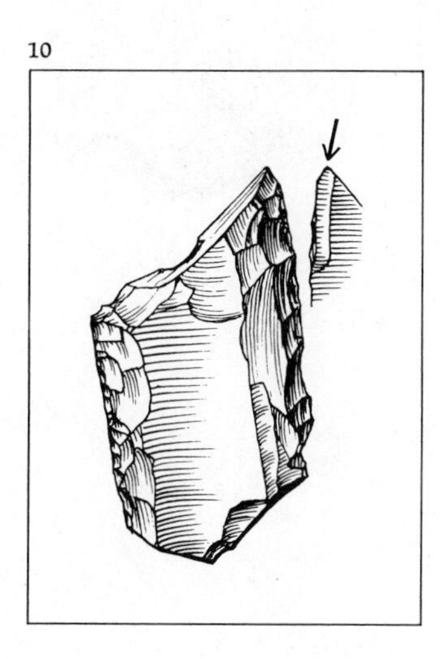
10

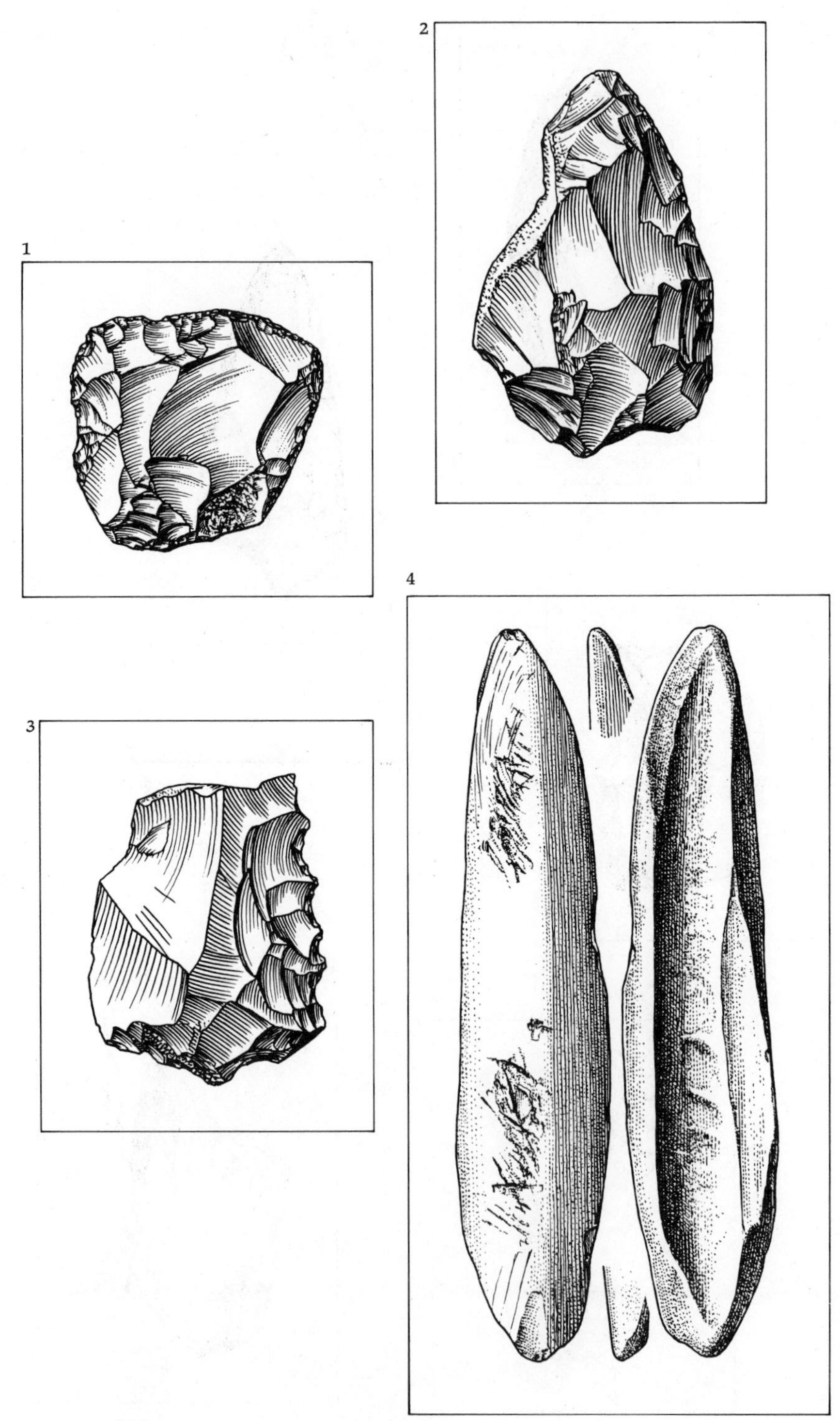
1
2
3
4

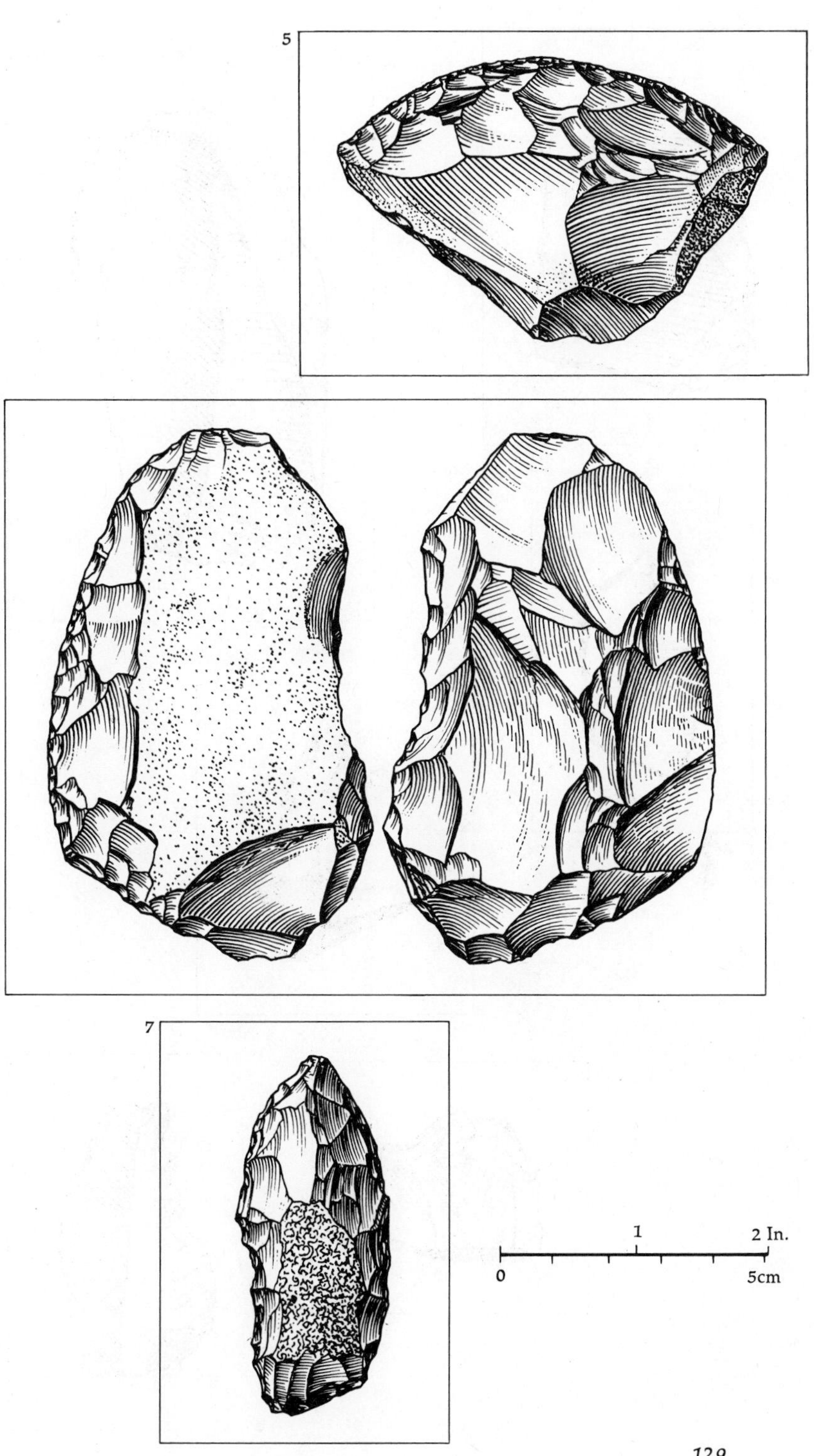
5
6
7
1
2 In.
0
5cm

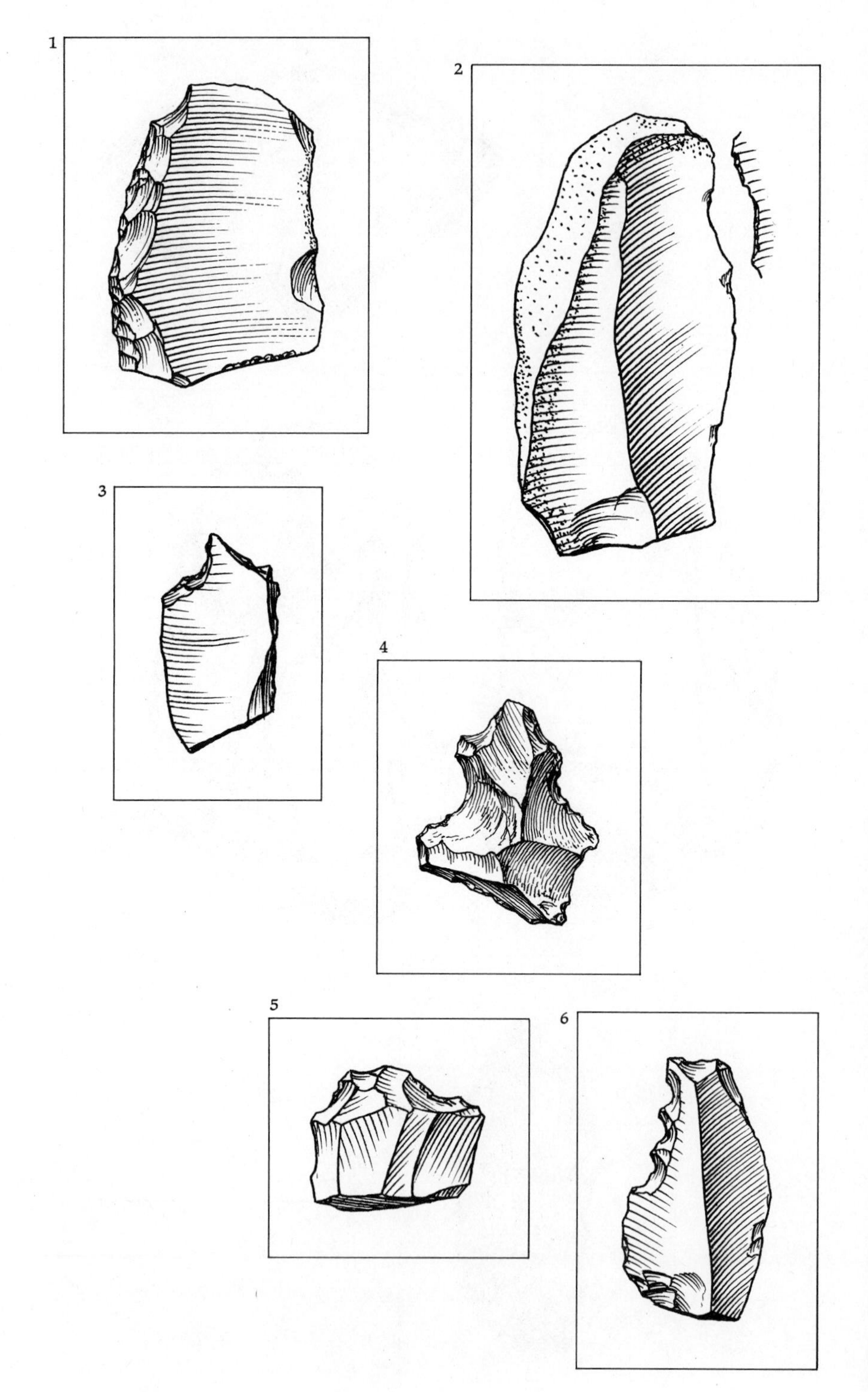
1
2
3
4
5
6

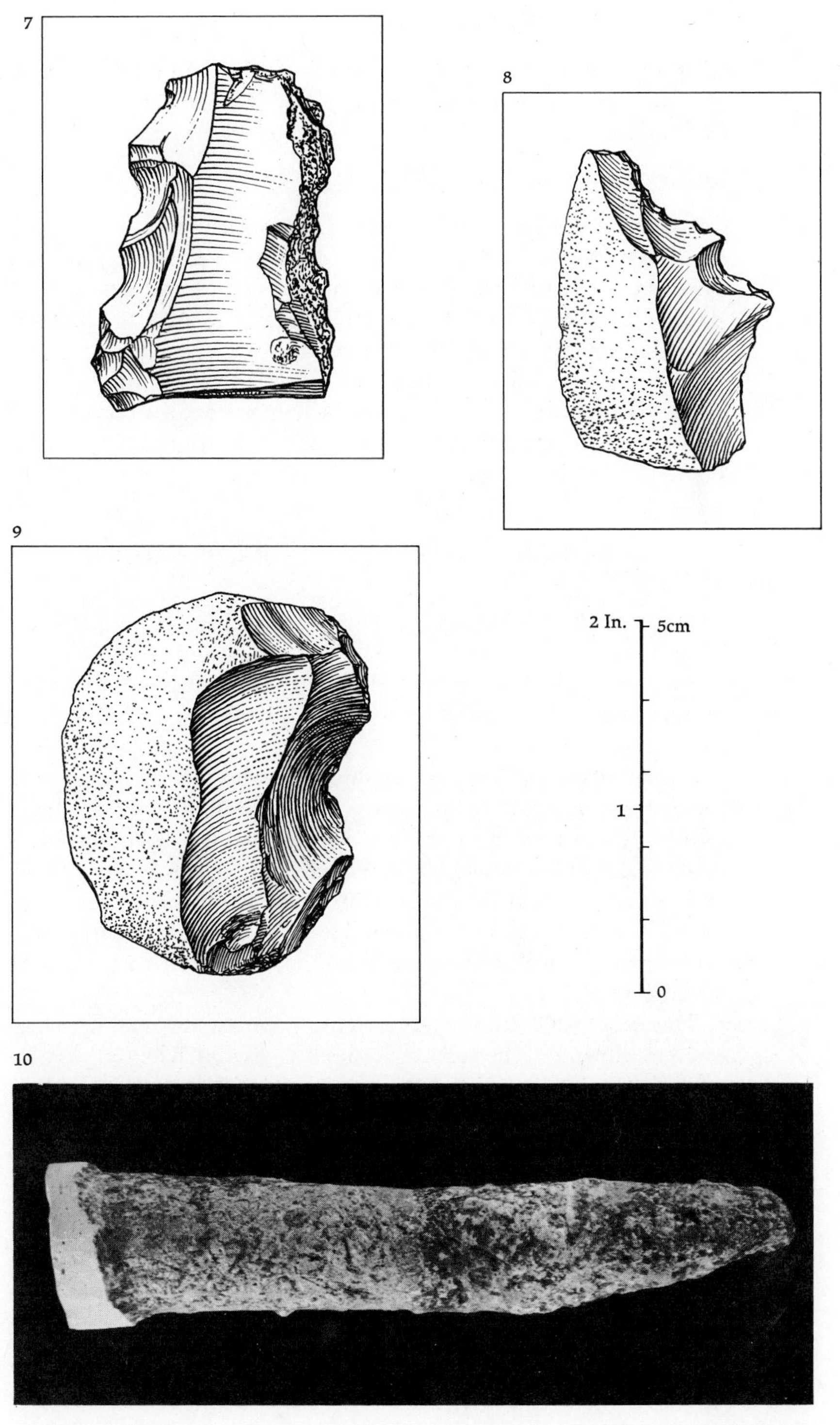
7
8
9
2 In.
5cm
1
0
10

Mousterian of Acheulean Tradition

This last type of Mousterian is poorly but clearly represented in the final layers of the sequence (Fig. 40).

RADIOCARBON DATES

For layer 12 we have radiocarbon dates, measured at Groningen in the Netherlands. Two samples have been dated. GrN-4304, taken in 1959, was composed of burned bones and has given a date of 39,000 ± 1500 B.P., that is, about 37,000 ± 1500 B.C. This date seems acceptable. Above layer 12, the layers are rich in large blocks and could have accumulated rather quickly.

The second date was done on hearth material, in 1964, and has given a date which seems much too young: GrN-4311, 30,300 ± 350 B.P.

PALETHNOLOGICAL OBSERVATIONS ON THE MOUSTERIANS AT COMBE-GRENAL

Most of the layers have not been yet studied for palethnological observations, but from what has been done and from observations during excavation, it is clear that concentrations of types of tools occur in several layers. Two interesting features have been found.

On August 5, 1959, Eugène Bonifay, who was working with me at the time, noticed in square F8, at the limit between layer 14 (gray) and layer 15 (yellow), a grayish circle of about 3 or 4 centimeters in diameter. He immediately thought of a posthole, and began to excavate it very carefully. It quickly became apparent that this was a cylindrical hole, and the posthole hypothesis became more and more likely. We did a plaster cast, in three sections (Fig. 39).

This cast is 206 millimeters in length, with a maximal diameter of 38 millimeters. The section is roughly circular. The tip, slightly askew, is pointed, and the end, which penetrated to layer 21, is blunted. The wooden post, hard-driven into the ground, met stone there and "mushroomed" a little. During the excavation,

→

Fig. 40

Combe-Grenal. Mousterian of Acheulean tradition.

1, Partial handaxe. 2, Backed knife. 3, Handaxe. 4, Backed knife. (1, 3, 4, layer 1. 2, layer 4).

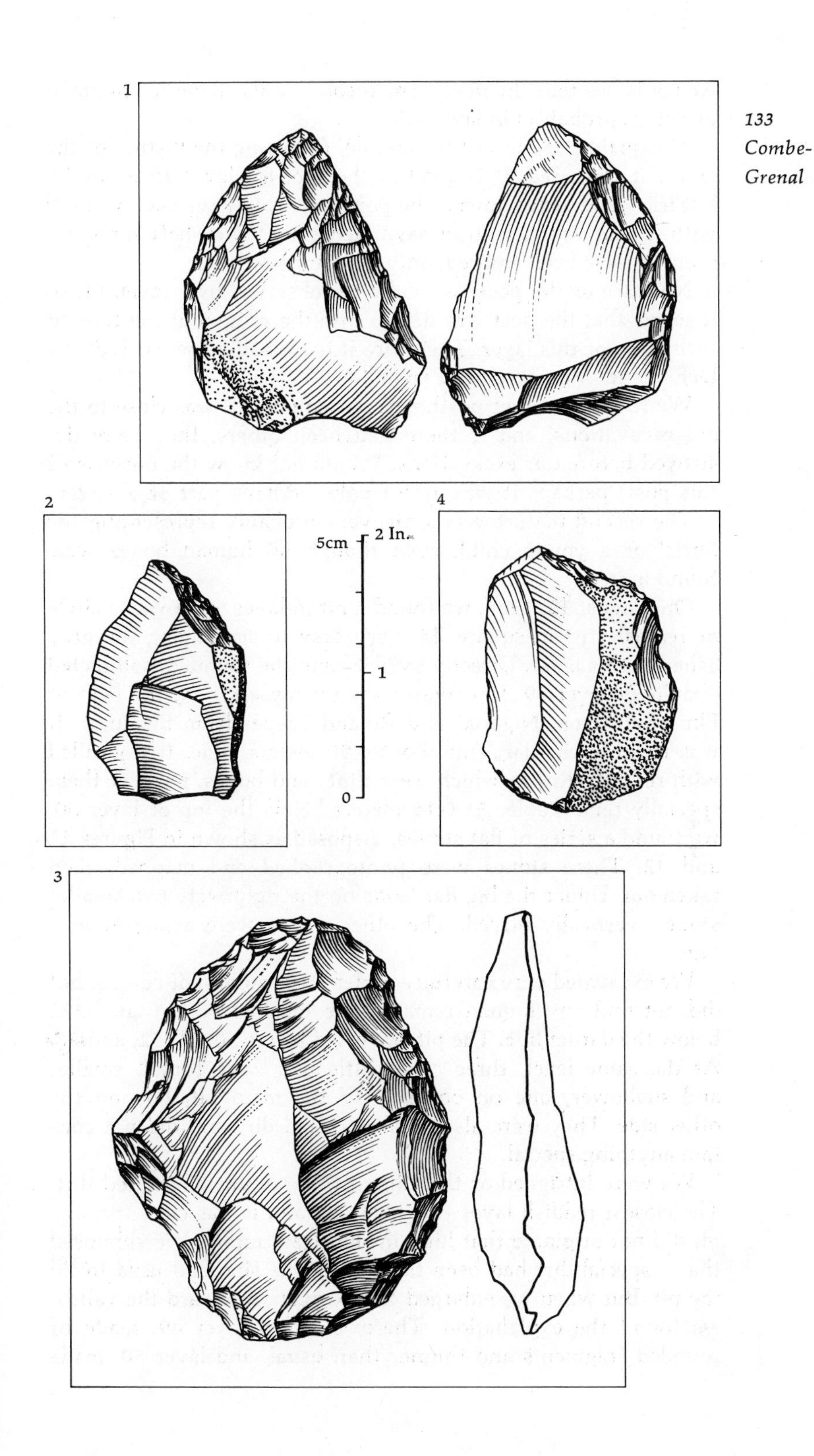

133
Combe-Grenal

we could see that the post went through a flat bone (a fragment of pelvis, probably) in layer 15.

The quality of the cast is variable, following the texture of the layers it traversed. It is good at the top, for layer 15 is sandy, but less good lower down. The point seems to have been worked with a flint knife, a little asymmetrically. The shaft does not seem to have been worked, only the bark taken off.

No trace of the posthole had been observed over layer 14, so it seems that the post was driven into the ground at the time of formation of this layer. Therefore it is probably the work of the Denticulate Mousterians of layer 14.

We found no other postholes, but square K8 was close to the old excavations, and if there had been others, they were destroyed before our excavations. We do not know the purpose of this post; perhaps it was a tent pole, perhaps part of a screen.

The second feature was a pit, very probably representing the burial of a young child, even though no human bones were found in it.

On August 12, 1961, we found a pit in layer 50. It was a circle of reddish dirt in square Z4, very easy to see among the gray ashes of this layer. Directly over it were the rounded, soliflucted eboulis of layer 49. I excavated the pit myself, with the help of Ekpo Eyo (from Nigeria) and Roland Paepe (from Belgium). It was roughly circular, and about 0.90 meters wide. It was filled with reddish dirt, in which were flints and bones, none of them specially remarkable. At 0.15 meters below the top of layer 50, we found a series of flat stones, disposed as shown in Figures 41 and 42. These stones were photographed and mapped, then taken out. Under the big flat stone on the right were two smaller stones, vertically placed. The other stones were arranged in a line.

We excavated very carefully under these stones, of course, but did not find any human remains; we touched bottom at −322 below the datum line. The pit traversed layers 50, 51, 52, and 53. At the same level, three other little pits were found, smaller and shallower; one on one side of the main pit, two on the other side. They were also filled with red dirt, but did not contain anything special.

We were intrigued at that time by the origin of the red dirt. The closest reddish layer was layer 42, and it was clear that the pit did not originate that high in the stratigraphy. We supposed that a special dirt had been brought to the site, and used to fill the pit. But when we enlarged the excavation toward the valley, we found the explanation. There, between layer 49, made of rounded fragments and thinner than usual, and layer 50, made

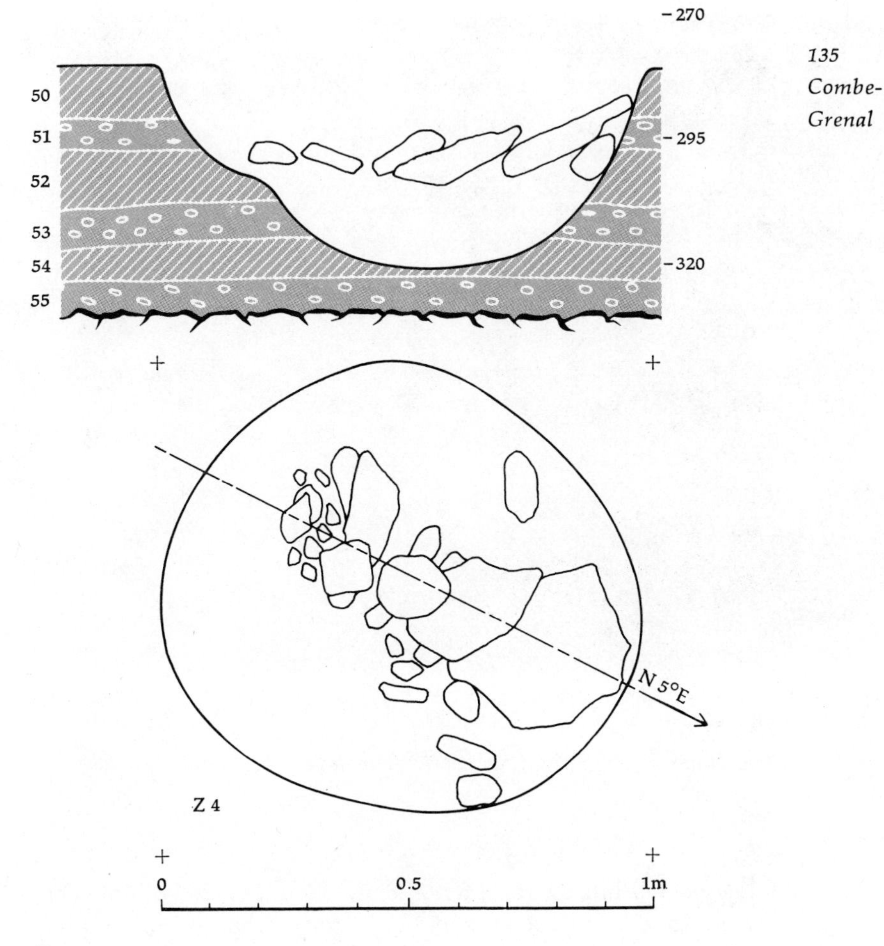

Fig. 41

Section and plan of the funeral pit in Typical Mousterian (Würm I) at Combe-Grenal, square Z-4.

of gray ashes, was intercalated a layer of reddish dirt, which we called 50A. It became clear that the pit had been deeper than its present state, and had been dug not from layer 50, but at a time when layer 50A was present there and at least partly deposited. The solifluctions of layer 49 afterwards swept this out in most places.

Why do we suppose it was a funeral pit, since we did not find human bones in it? We know of other occurrences of funeral pits in Mousterian sites which look very much the same and

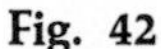
Fig. 42

The funeral pit in the lower Würmian layers at Combe-Grenal.

have yielded human remains. A child sepulcher at Le Moustier, for instance, was excavated by D. Peyrony, who says in his 1930 paper:

> Near the part preserved to show the stratigraphy of the site was a pit, in the shape of a truncated cone, dug into the sandy layer I and the underlying layer H. Its upper diameters were respectively 0.70 meters and 0.80 meters, and it was 0.60 meters deep. It was filled with broken bones and the dirt of the layer I have just described [layer J], with some fine Mousterian tools.
>
> It was covered with three juxtaposed stones, put there intentionally after the pit had been filled up.
>
> Southwest of this pit, and about 1 meter away, I found another, 0.50 m wide and 0.40 m deep, dug in the same conditions. It contained the skeleton of a young child. . . .
>
> Since the two pits were dug at the same period, one can

posit a relationship between them. Would it be foolish to suppose that the first could have been dug to receive the remains of a funeral meal, or maybe the offerings, food and weapons, which would then have been covered to protect them? (Fig. 43) (27)

It is also possible that the first pit at Le Moustier contained the corpse of a child so young that his bones had not fossilized. In the case of Combe-Grenal this hypothesis is reinforced by the fact that layers 51 and 53, made only of rounded eboulis, without dirt, are very permeable and serve as an evacuation route for the water percolating into the ground. Such circumstances would not be very favorable to the conservation of the bones of a very young child.

This association of several pits has also been found at La Chapelle-aux-Saints and La Ferrassie, in southwest France.

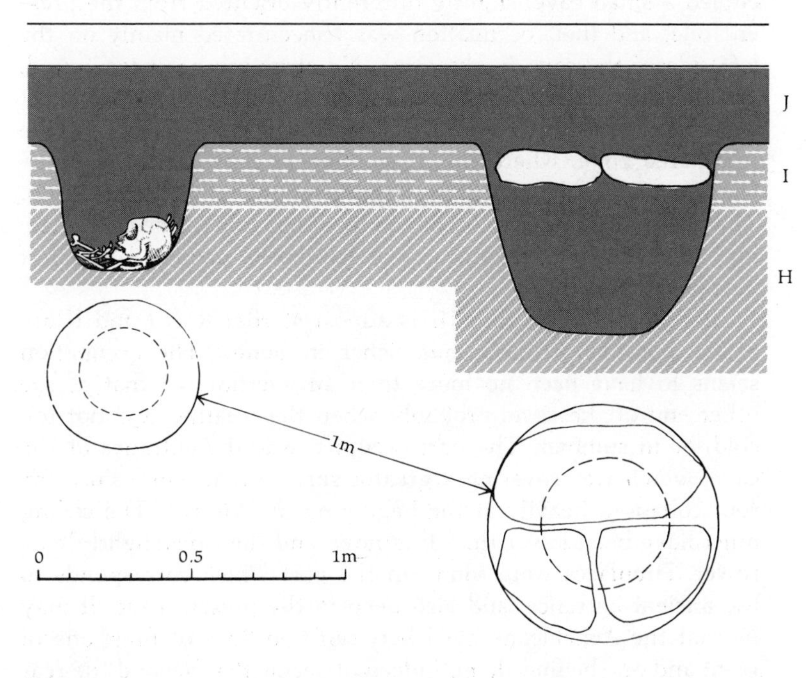

Fig. 43

Section and plan of the funeral pits in Typical Mousterian (Würm II) at Le Moustier, after Peyrony, 1930 (slightly modified).

5. *Pech de l'Azé and Combe-Grenal Compared*

Together, these two sites cover the whole sequence from the end of the Mindel-Riss interglacial to Würm II, with the exception of the Riss-Würm interglacial, marked only by weathering and soil formation (see Table 2).

Riss I and II are not represented at Combe-Grenal. Perhaps they existed once, more to the front of the site, but if so they have been eroded away. Riss III layers have given a rich meridional Upper Acheulean, as we have seen. The Acheuleans occupied a small cave, slightly differently oriented from the present one, and their occupation was concentrated mainly on the left side of the cave (looking in). No structure has been found, except for small fireplaces without any special preparation; but it should be stressed that the richest layers were mainly in the weathered zone, where the dissolution of most of the congelifracts certainly disturbed the arrangement of tools, and so on. The horizontal distribution of the tools has not yet been studied, but there was a concentration of handaxes, mainly in the left part of the cave.

At Pech de l'Azé, Riss III is almost sterile; Riss I and II are rather poor in artifacts, but richer in bones. The occupation seems to have been no more than an overflow of that at the other end of the cave, probably when the weather was not too cold, or in summer. The part occupied was the entrance of the cave, which was covering a greater surface than now, since the roof collapsed heavily at the beginning of Würm I. The ceiling must have been lower than it is now, and the cave slightly narrower. Fireplaces were found in the part which corresponds to the ancient entrance, and also deep in the present cave. It may be that the Acheuleans lived between two lines of fires, one in front and one behind them. Indeed, it seems that beyond the rear line of fires, almost no tools have been found.

The beginning of Würm I at Combe-Grenal is marked by a cold and damp period, which does not seem to be represented at Pech de l'Azé II. This is layer 55, heavily cryoturbated. It was followed by a cold dry period, also with Typical Mousterian. This is probably the period during which the first Typical Mousterians (4E) at Pech de l'Azé inhabited the cave. Then, in both sites, we have a temperate period; but whereas at Combe-Grenal this period is still occupied by Typical Mousterian, in Pech II we find first Typical Mousterian, then Denticulate Mousterian, then Typical Mousterian again. It was probably the onset of this warmer period, with alternate thawing and freezing, that was responsible for the cryoturbation of layers 53 to 55 at Combe-Grenal and for the polygonal soil and the cryoturbation of layer 4D at Pech de l'Azé. During this temperate period, at Combe-Grenal we have first a layer rich in black ashes (52), then another cryoturbated layer (51), then a layer very rich in white ashes (50), then a layer of reddish sands (50A). At Pech de l'Azé II, there is a reddish sand, probably deriving its color from older formations, even if there is a slight pedogenesis.

Then comes another cold period, with Typical Mousterian in both sites. But if the cryoturbations are slight in Pech de l'Azé layer 3, they are strong at Combe-Grenal, and most of the implements are battered (layers 49 to 44).

During the second warmer period, we have Typical Mousterian (42 and 43) and perhaps Denticulate (41) in Combe-Grenal; but if 2G′ is, possibly, still Typical Mousterian at Pech de l'Azé, 2G is already Quina-Ferrassie type. This warm period is cut in two by a short cold spell, marked by pollen analysis at Pech de l'Azé, and by a thin layer of congelifracts at Combe-Grenal. The difference in the percentage of tree pollens between the two sites is also notable: 33 percent at Pech de l'Azé, 60 percent at Combe-Grenal; but in both cases oak is present. Even today the small valley of Pech de l'Azé is colder than the Dordogne Valley, which Combe-Grenal is near.

This warm period is followed by a dry cold one, with Typical Mousterian at Combe-Grenal and Quina-Ferrassie at Pech de l'Azé, where the sequence is almost finished.

At Combe-Grenal comes a third warmer period, although less warm than the preceding ones, corresponding perhaps in Pech II to layers 2A to 2C. In the first site this period is occupied by Denticulate Mousterian, with Levallois-flaking technique.

Würm I ends with a last cold period, with Typical Mousterian. The interstadial is marked at Combe-Grenal by the development of a soil on top of the Würm I layers, and at Pech de l'Azé by a very strong erosion, emptying Pech I almost completely.

Würm II begins in both sites with a very cold and damp period. But while at Pech de l'Azé I we have Mousterian of Acheulean tradition (layers 3 and 4), at Combe-Grenal we have

Table 2

	PECH DE L'AZÉ					
	CLIMATE	% TREES	FAUNA	MOUSTERIAN	LAYER	CLIMATIC ZONES
PECH I				Sterile		VIII
						VII
						VI
						V
				Acheulean Tradition, Type B	7	IV
					6?	III
	Cold, dry	5	Bovids, Reindeer	Acheul. Trad. Type B	5	II
	Very cold, damp	4–6	Bovids, Red deer, Reindeer	Acheulean Tradition, Type A	4	I
			EROSION			
					1?	VII
			Red deer		2A–2C	VI
		<10	Red deer, Bovids		2D	
	Cold ↑, dry ↑	10	*Reindeer*, Red deer	Ferrassie	2E	V
		12	Reindeer, Bovids, Red deer		2F	

Continued on page 142

COMBE-GRENAL				
CLIMATE	% TREES	FAUNA	MOUSTERIAN	LAYER
Very cold, dry	<10	Red deer, Reindeer, Bovids	Acheulean Tradition	1–4 5–6
Less cold, damper	15–18	Red deer, Reindeer	Typical Typical	7 8
Colder, drier	<10	Bovids, Red deer, Reindeer	? Typical	9 10
A little less cold, damp	15–17	Bovids, Reindeer Bovids, Horses Bovids, Horses	Denticulate Denticulate Denticulate	11 12 13
Very cold, steppic	5	Horse, Reindeer Reindeer, Horse	Denticulate Quina	14–16 17–19
A little less cold, damper	14–16	Reindeer, Horse	Denticulate Quina Quina	20 21 22
Cold, dry, steppic	5	Reindeer, Red deer	Quina	23–25
Very cold, damp	5–7	Red deer, Reindeer	Quina Ferrassie Typical Ferrassie	26 27 28–31 32–35
SOIL WÜRM I/WÜRM II				
Cold, very dry	11	Red deer	Typical	36–37
MILDER, damp	33	Red deer	Denticulate	38
Cold, dry	15 13	Red deer Reindeer	Typical? Typical	39 40

Continued on page 143

Table 2 (cont.)

	PECH DE L'AZÉ					
	CLIMATE	% TREES	FAUNA	MOUSTERIAN	LAYER	CLIMATIC ZONES
	EROSION					
PECH II	MILD (OAK)	33	*Red deer*, Reindeer	Ferrassie	2G	IV
	Cold, dry	10		Ferrassie	2G base 2G' top	
	MILDER, damp	23	*Red deer*, Reindeer	?	2G'	
	Cold dry / damp	8 / 17	Red deer, Reindeer	Typical	3	III
	MILD, DAMP	57	Red deer, Reindeer	Typical	4A	II
		50	Red deer, *Horse*, Reindeer	Denticulate	4B	
		43	Red deer, Horse, Bovids	Typical	4C	
	Cold, dry	10	Red deer, Reindeer	Typical	4D 4E	I

a succession of different types: Ferrassie (layers 36 to 32), Typical (layers 31 to 28), Ferrassie again (layer 37), and Quina (layer 26). This damp period is followed by a cold, dry, steppic period, with Mousterian of Acheulean tradition (layer 5) at Pech I and Quina (layers 25 to 23) at Combe-Grenal.

The third oscillation, less cold and damper, corresponds probably (the analysis has not yet been done) to layer 6 at Pech de l'Azé (Mousterian of Acheulean tradition) and at Combe-Grenal consists of Quina Mousterian (layers 22 and 21) and Denticulate Mousterian (layer 20). The following period, very cold and

COMBE-GRENAL				
CLIMATE	% TREES	FAUNA	MOUSTERIAN	LAYER
SOIL WÜRM I/WÜRM II				
MILD (OAK)	60	Red deer	Denticulate?	41
←Thermoclastic Eboulis→				
MILD (OAK)		Red deer	Typical	42–43
Dry, cold	14 24 ↑	Red deer, WILD BOAR	? Typical ?	44–45 46–47 48–49
MILD	60	Red deer, WILD BOAR [Reindeer in 52]	Typical	50A–52
Cold, dry	12	Red deer, Wild boar	Typical	53–54
Cold, damp	15			55

steppic, is probably coeval with layer 7 at Pech I (still Mousterian of Acheulean tradition), but consists of the last Quina layers (19 to 17) and Denticulate Mousterian (layers 16 to 14) at Combe-Grenal.

The fifth oscillation, less cold but damp, corresponds at Pech I with almost sterile layers, and at Combe-Grenal with Denticulate Mousterian (13 to 11). The sixth, colder, has Typical Mousterian (layer 10) and indeterminate Mousterian (layer 9) at Combe-Grenal; the seventh oscillation, less cold but damper, begins at Combe-Grenal with Denticulate Mousterian (layer 8), then goes on with Typical Mousterian (layer 7). The last period of Würm II, very cold, sees Typical Mousterian again (layer 6), then Mousterian of Acheulean tradition.

So far as Pech de l'Azé I is concerned, after layer 5 the correlations are hypothetical, since the analysis is only now under way. It may be that we find the equivalent there of the whole Combe-Grenal sequence.

From all this, we can assume that climatic and typological variations are not linked. Any kind of Mousterian can be found under widely different climatic conditions. The typological variations are not linked with faunal variations either, even if the Denticulate Mousterian seems to have hunted horse by predilection whenever possible.

6. Differences Between Würm I and II Faunas

In the southwest of France, Würm I faunas are characterized by the predominance of the red deer. It is a small-sized variety, or subspecies. Usually, it is accompanied by other forest species, such as the roe deer and wild boar. The presence of reindeer is accidental, except toward the end. Mountain goats are absent. Present but never very abundant are the ass (*Equus hydruntinus*), the big Irish deer (*Cervus megaceros*), and Merck's rhinoceros. Bovids and horses are never very abundant.

During Würm II, reindeer plays a much more important role and becomes the best represented species in many layers. Only horses and bovids, toward the end of the period, and red deer, toward the beginning, sometimes take precedence. This red deer is definitely bigger than the one in Würm I. Mountain goat is sometimes abundant. Merck's rhinoceros is replaced by the woolly rhinoceros.

7. *The Significance of the Different Mousterian Industries*

The different Mousterian industries seem well established today, and at Combe-Grenal or Pech de l'Azé we have encountered roughly all those which exist in western Europe. Others exist elsewhere, in central or eastern Europe, for instance. But what is the significance of this variability? We tend to interpret these different industries as reflecting the cultural differences of human groups in possession of varied traditions. Others prefer to explain these variations as the result of different activities carried on by people of the same culture. And others again think that the different Mousterians represent different steps in the evolution of the Mousterian culture. I have recently published a paper on this subject (15), but it is may be interesting, at the close of this tale of two caves, to give a brief summary of the present state of the question.

Our point of view is that during Mousterian times different cultures, with different traditions of tool-making and tool-using, coexisted on the same territory but influenced each other very little. Several criticisms can be offered of this point of view. One is that cultures do not last that long, to which one can reply that something did last for millennia (with some changes, of course), and that is the assemblages. Another criticism is that it is unlikely that different contemporary cultures would not influence each other, whether by marriage or contact, or that they could exploit the same ecological niche in the same territory.

Here we should examine what is meant by such terms as "contemporary" and "same territory." One should never forget the imprecision of our chronology and the length of prehistoric times. If in two shelters, A and B, which are close to each other, different Mousterians are "contemporary" within the limits of our possibilities of dating, it may well be that 100 or more years elapsed between the abandonment of shelter A by Mousterians of type X and the occupation of shelter B by Mousterians of type Y. Sometimes, however, a real contemporaneity seems to have been possible. At the beginning of Würm I, as we have seen, Typical Mousterians occupied Combe-Grenal, while Typical, then Denticulate, then Typical again occupied Pech de l'Azé II. So the Denticulate Mousterians of Pech de l'Azé should be the contemporaries, strictly speaking, of one or the other of the Typical Mousterians of Combe-Grenal. The two shelters are close, but separated by the Dordogne River, which may well have been the frontier.

Intermarriages are difficult to assert or refute, but in primitive societies, conservatism is usually very strong. If one supposes that a Mousterian of Acheulean tradition man married a Quina woman, she might have gone on using the thick scrapers to which she was accustomed, but we doubt that her daughters would have done the same.

The hypothesis that the different Mousterian types represent an evolution of the same general Mousterian culture is negated by the numerous interstratifications known today. We have seen that climatic influences do not seem to be any better explanation. The "different activities" hypothesis, outlined by Lewis and Sally Binford (1, 2), needs closer examination. First, let us get rid of one variant, the seasonal activities theory. Each type of Mousterian assemblage would correspond to one season. But the layers are often thick, without any significant change from top to bottom, and must have accumulated over a long period of time. We should then presuppose a covenant between tribes, reserving such and such a cave for summer activities, and others for different seasonal activities. Moreover, we know that in some caves, at least, man was present all the year round.

The Binfords' point of view is different. Using the results of Sally Binford's excavation in Israel, and some of my own excavations, the authors have submitted the count of tool types to factor analysis and obtained several factors which they interpret as representing specialized activities or activity complexes. They deduce the existence of different types of sites: living sites, hunting sites, workshops, and so on, and thus explain the different types of Mousterian as representing different activities, more or less predominant, following the specialization of the site.

Even if we accept the validity of factor analysis as used by

Lewis Binford (some criticisms have been presented on purely methodological grounds), several objections can be raised against this theory. Killing sites and workshops, as opposed to what happened in America, were rare. Usually, tools were made and used in the same place. Even if the factors exist, their meaning derives from an interpretation always open to criticism. The computer tells us that such and such tools covary; it does not say for what they were used.

Again, ethnographic comparisons may well be dangerous if they do not take into account the broad differences of environment. Southwest France never was the Kalahari, the Australian desert, or Labrador. The ways of life may well have been quite different. Flint is plentiful almost everywhere, so no specialized quarries were needed, nor trade favorable to contact, as in Australia.

We do know of specialized tool kits, corresponding probably to different activities, but within a single site. They probably show that different activities were carried on at different places but always within the same site.

If the Binfords' theory is correct, then we would expect open-air sites to give different assemblages from those of the caves. But this does not seem to be the case, and the differences, when they exist, are too small to transform one type of Mousterian into another.

One can raise the same objection to the hypothesis of different activities as to the hypothesis of seasonal activities. Some layers, accumulated over a long time, do not show much internal typological variation. Here again, should we suppose a covenant among Mousterian people to reserve such-and-such a cave for such and such an activity, or complex of activities? If not, one would expect a continuous spectrum of variability in the assemblages, which is clearly not so; the variations are discrete. Furthermore, the different Mousterian assemblages clearly show differences in style. The thin scrapers of the Quina Mousterian are different from the thin scrapers of the Mousterian of Acheulean tradition, and so on.

Some territories have been occupied for a very long time by the same kind of Mousterians. In Charente, the Mousterian of Acheulean tradition is rare, while the Charentian (Quina-Ferrassie) seems to be virtual master. The Combe-Grenal site seems to have been occupied during most of Würm I by Typical Mousterian. In Provence, the Mousterian of Acheulean tradition is unknown. One cannot help wondering what kind of activities were undertaken in the Dordogne under this facies, which were unnecessary in Provence. And if the answer is that the same activities were being performed in a different way, then this means that there are different ways of performing the same activities with different tool kits. So why not admit that the different Mousterian types just represent these different ways, and that the difference is indeed cultural?

8. The Work Goes On

Perhaps this tale of two caves has proved a disappointment to you. Perhaps it contained too much cold science and not enough about the human life of those remote times. But, unless I wish to write a science fiction story, it is not possible to go further. Most of the paleethnological work has still to be done; the relevant information still lies hidden in the notebooks. We have been busy establishing the chronological and typological framework. In a few years, I hope, I will be able to say more. But even if all the work had already been completed for Pech de l'Azé and Combe-Grenal, we would need more caves excavated in the same way, or even more thoroughly, before we could tell for sure what is chance and what is really significant. More than any other science, prehistory is a game of patience. We accumulate facts, then build up hypotheses and try to prove or disprove them. It is not easy. Experimental prehistory is, perforce, limited to the reconstruction of prehistoric tools and their use. But we are handicapped by the fact that we are *not* prehistoric men. We know too much, and not enough. We know the final results, we know that such-and-such a type of tool is possible, since we have found it; but we cannot know for sure what steps were taken by paleolithic man in its invention. We re-create, but we cannot create.

And so the work goes on, slowly. Probably not slowly enough. It took me 13 years to excavate Combe-Grenal. If I had to do it again, with what I now know of what I should have done, it would probably take me 20 years. Prehistory is a great teacher of patience and humility. To excavate is to destroy, and, as a friend of mine once put it: "There is no such thing as a good excavation, there are only some that are not as bad as others."

Excavating must go on, since it is the only way really to learn. But it is sacred work that should not be taken lightly. We are disturbing the dead, sometimes violating sepulchers, and our only excuse is this hunger to know what separates man from the animals.

Yes, the work goes on. We are now excavating Pech de l'Azé IV, a collapsed shelter about 100 meters east of Pech I, which will probably give us more information on Würm I. To do something seriously does not mean it must be done gloomily. There is a good deal of fun and laughter at the excavation. People from different countries are united in the task; this year, we had eight different nationalities and three different races. The origin of man, the way he lived so long ago, and the history of his efforts to master the natural environment interest any human being, whatever his race, age, sex, creed, or nation. The work will go on for a long time. After Pech de l'Azé IV (and more teams from the Laboratory are excavating other sites), there will be other caves, other shelters. And other archaeologists. Some day, why not you?

Montreal-Bordeaux. Fall, 1970

Appendix: Typological and Technological Counts for Layer 4, Pech de l'Azé I

(MOUSTERIAN OF ACHEULEAN TRADITION, TYPE A)

		REAL COUNT	
	NUMBER	%	% CUMULATED
1. *Typical Levallois flakes*	36	0.91	
2. *Atypical Levallois flakes*	135	3.44	4.35
3. *Levallois points*	5	0.12	4.47
4. *Retouched Levallois points*	0	0	4.47
5. *Pseudo-Levallois points*	37	0.94	5.41
6. *Mousterian points*	20	0.51	5.92
7. *Elongated Mousterian points*	3	0.07	5.99
8. *Limaces*	1	0.02	6.01
9. *Single straight side scrapers*	142	3.62	9.63
10. *Single convex side scrapers*	257	6.55	16.18
11. *Single concave side scrapers*	61	1.55	17.73
12. *Double straight side scrapers*	8	0.20	17.93
13. *Double straight-convex side scrapers*	23	0.58	18.51
14. *Double straight-concave side scrapers*	1	0.02	18.53
15. *Double convex side scrapers*	23	0.58	19.11
16. *Double concave side scrapers*	2	0.05	19.16
17. *Double concave-convex side scrapers*	8	0.20	19.36
18. *Convergent straight scrapers*	6	0.15	19.51
19. *Convergent convex scrapers*	30	0.76	20.27
20. *Convergent concave scrapers*	0	0	20.27
21. Déjeté *(offset) scrapers*	37	0.94	21.21
22. *Straight transverse scrapers*	7	0.17	21.38
23. *Convex transverse scrapers*	18	0.45	21.83
24. *Concave transverse scrapers*	2	0.05	21.88
25. *Side scraper on ventral face*	154	3.93	25.81
26. *Abrupt retouched side scrapers*	22	0.56	26.37
27. *Side scrapers with thinned back*	6	0.15	26.52
28. *Side scrapers with bifacial retouch*	27	0.68	27.20

	ESSENTIAL COUNT	
NUMBER	%	% CUMULATED
	0	0
	1.47	1.47
	0.79	2.26
	0.12	2.38
	0.04	2.42
	5.63	8.05
	10.20	18.25
	2.42	20.67
	0.31	20.98
	0.91	21.89
	0.04	21.93
	0.91	22.84
	0.08	22.92
	0.31	23.23
	0.23	23.46
	1.19	24.65
	0	24.65
	1.46	26.11
	0.27	26.38
	0.71	27.09
	0.08	27.17
	6.11	33.28
	0.87	34.15
	0.23	34.38
	1.07	35.41

Typological and Technological Counts for Layer 4, Pech de l'Azé I—continued

	NUMBER	REAL COUNT %	% CUMULATED
29. *Alternate retouched side scrapers*	56	1.42	28.62
30. *Typical end scrapers*	33	0.84	29.46
31. *Atypical end scrapers*	32	0.81	30.27
32. *Typical burins*	20	0.51	30.78
33. *Atypical burins*	38	0.96	31.74
34. *Typical borers*	23	0.58	32.32
35. *Atypical borers*	33	0.84	33.16
36. *Typical backed knives*	15	0.38	33.54
37. *Atypical backed knives*	28	0.71	34.25
38. *Naturally backed knives*	92	2.34	36.59
39. *Raclettes*	163	4.16	40.75
40. *Truncated blades and flakes*	48	1.22	41.97
41. *Mousterian tranchet*	6	0.15	42.12
42. *Notches*	264	6.73	48.85
43. *Denticulates*	459	11.71	60.56
44. *Alternate retouched beaks*	39	0.99	61.15
45. *Retouches on ventral face*	102	2.60	64.15
46–47. *Abrupt and alternate retouch (thick)*	127	3.24	67.39
48–49. *Abrupt and alternate retouch (thin)*	973	24.83	92.22
50. *Bifacial retouch*	22	0.56	92.78
51. *Tayac points*	19	0.48	93.26
52. *Notched triangles*	21	0.53	93.79
53. *Pseudo-microburins*	4	0.10	93.89
54. *End-notched pieces*	32	0.81	94.70
55. *Hachoirs*	10	0.25	94.95
56. Rabots *(pushplanes)*	0	0	94.95
57. *Tanged points*	0	0	94.95
58. *Tanged tools*	1	0.02	94.97
59. *Choppers*	0	0	94.97
60. *Inverse choppers*	0	0	94.97
61. *Chopping tools*	1	0.02	94.99
62. *Miscellaneous*	184	4.69	99.68
63. *Bifacial leaf-shaped points*	1	0.02	99.70

Total of tools, in real count: 3,918; in essential count: 2,518. Ordinary flakes: 3,522. Ordinary blades: 314. Handaxe flakes: 24,875. Chips, etc.: about 700.

	ESSENTIAL COUNT	
NUMBER	%	% CUMULATED
	2.22	37.67
	1.31	38.98
	1.21	40.25
	0.79	41.04
	1.50	42.54
	0.91	43.45
	1.31	44.76
	0.59	45.35
	1.11	46.46
	3.65	50.11
	6.47	56.58
	1.90	58.48
	0.23	58.71
	10.48	69.19
	18.22	87.41
	1.54	88.95
	0.75	89.70
	0.83	90.53
	0.16	90.69
	1.27	91.96
	0.39	92.35
	0	92.35
	0	92.35
	0.04	92.39
	0	92.39
	0	92.39
	0.04	92.43
	7.30	99.73
	0.04	99.77

Typological and Technological Counts for Layer 4, Pech de l'Azé I—continued

	HANDAXES		
	N	%	% CUMULATED
1. *Lanceolates*	2	1.28	
2. *Ficrons*	0	0	1.28
3. *Micoquians*	0	0	1.28
4. *Triangular handaxes*	7	4.48	5.76
5. *Elongated triangular*	0	0	5.76
6. *Cordiform handaxes*	55	35.25	41.01
7. *Elongated cordiform*	5	3.20	42.21
8. *Subcordiform*	31	19.87	64.08
9. *Oval-shaped*	2	1.28	65.36
10. *Amygdaloid*	0	0	65.36
11. *Discoid*	11	7.05	72.41
12. *Limandes (ellipsoid, flat)*	0	0	72.41
13. *Bifacial cleavers*	1	0.64	73.05
14. *Flake cleavers*	0	0	73.05
15. *Lageniform (bottle-shaped)*	0	0	73.05
16. *Lozenge-shaped*	0	0	73.05
17. *Naviform*	0	0	73.05
18. *Nucleiform*	9	5.76	78.81
19. *Miscellaneous*	12	7.69	86.50
20. *Partial*	21	13.46	99.96
21. *Abbevillian*	0	0	99.96

Total of handaxes: 156. Fragments: 38.

DISKS: 8. PICKS: 0. HAMMERSTONES: 0.

Levallois

	PLAIN	FACETED	CONVEX-FACETED	DIHEDRAL-FACETED	TAKEN AWAY	BROKEN, ETC.
FLAKES	64	103	85	40	16	108
POINTS	1	4	3	3	0	0
BLADES	23	48	19	6	8	58

Total: 589

Non-Levallois

	PLAIN	FACETED	CONVEX-FACETED	DIHEDRAL-FACETED	TAKEN AWAY	BROKEN, ETC.
FLAKES	1,414	615	463	692	261	2,297
POINTS	0	2	8	27	0	1
BLADES	222	94	43	64	16	192

Total: 6,411

Grand total: 7,000

QUARTZ:
Flakes: 272. Fragments: 442. Pebbles: 11

BASALT:
Flakes: 123. Fragments: 135. Pebbles: 26

OTHER ROCKS:
Pebbles: 34

MANGANESE DIOXIDE:
Lumps: 78. Pencils: 124. Scratched lumps: 47

YELLOW OCHER:
Lumps: 4

RED OCHER:
Lumps: 23

PALETTE: 1

UTILIZED BONES: 4

CORES:
Levallois for flakes: 18. Levallois for points: 1
Levallois for blades: 0. Discoid: 79. Globular: 33
Prismatic for blades: 4. Pyramidal: 3. Miscellaneous: 80
Shapeless: 57

INDEXES

Levallois index: 8.41. Faceting index: 57.35. Faceting index (restricted): 36.77. Laminary index: 11.32.

Real typological indexes and characteristic groups

Typological Levallois index: 4.47. Scraper index: 22.61. Unifacial Acheulean index: 1.09. Handaxe index: 3.79.

Levallois group (I): 4.47 Mousterian group (II): 24.15. Upper Palaeolithic group (III): 6.85. Denticulate group (IV): 11.71.

Essential typological indexes and characteristic groups

Typological Levallois index: 0. Scraper index: 35.25. Unifacial Acheulean index: 1.70. Handaxe index: 5.83.

Levallois group (I): 0. Mousterian group (II): 37.67. Upper Palaeolithic group (III): 10.69. Denticulate group (IV): 18.22.

Glossary

Abbevillian: *The oldest handaxe-bearing culture in Europe, characterized by rough handaxes worked with a stone hammer.*

Assemblage: *All the artifacts found in a given layer at a site.*

Aurignacian: *An industry characterized by thick (carinate) scrapers, retouched blades, and numerous bone tools; contemporary with the Perigordian.*

Azilian: *The first postglacial culture in western Europe representing an adaptation of the Magdalenians to changing conditions.*

Blade: *An elongated flake, at least twice as long as it is wide. Sometimes the term blade is restricted to parallel-sided elongated flakes.*

Breccia: *Consolidated sediment made of angular fragments cemented by lime or silica.*

Burin: *A tool with a cutting edge formed by the intersection of two small removals (burin spalls) or by the intersection of one such removal and a facetted retouch (burin on truncation).*

Clactonian: *A Lower Palaeolithic culture, located at Clacton on Sea on the east coast of England, without handaxes, but with choppers and chopping tools. It is thought to be a western extension of a great complex of cultures that did not have handaxes and that stretched from Eastern Asia to Europe.*

Core: *Nodule of flint, obsidian, and so on, from which flakes or blades have been detached.*

Cryoturbation: *Deformation of layers of soil caused by alternate conditions of freezing and thawing.*

Eolian: *Material deposited by wind action.*

Factor analysis: *A complex statistical technique used to analyze quantified data. It measures covariation among variables and dis-*

covers meaningful groupings among them. In archaeology, counts of tool types or other things have been subjected to a factor analysis in order to discover groupings of such tools that are consistently found together. Use of a computer is virtually essential to undertake a factor analysis.

Glacial, Glaciation: *A major cold period, comprising several stadials and interstadials. An interglacial is a long, warmer period between glacials.*

Günz: *The oldest of the classical glacial periods recognized in Europe. At least two older glacial periods (Biber and Donau) are now thought to exist.*

Industry: *A type of assemblage found time and again in a given region at a given time.*

In situ: Latin words meaning on the spot, undisturbed.

Levallois Technique: *Technique of preparing a core in such a way that the shape of the flake to be struck off is predetermined. It may comprise the preparation of the striking platform by removing small flakes producing facets.*

Limace: *French word meaning "slug"; name given by ancient prehistorians to ellipsoidal unifacial tools, recalling more or less a slug shape; found in the Mousterian.*

Magdalenian: *The last Pleistocene culture in western Europe; began about 16,500* B.C., *and ended about 9,500* B.C. *The Magdalenian is well-known for its art.*

Micoquian: *One of the branches of the Final Acheulean; partially contemporary with the Mousterian and characterized by finely worked lanceolate handaxes.*

Microlith: *Small stone tools made on flakes or blades, often less than one inch long, which may be geometrically shaped.*

Mindel: *The second of the classical glacial periods and perhaps the most important as far as the surface covered by the ice caps is concerned.*

Mousterian: *Palaeolithic culture coeval with the first half of the Würm glaciation; began about the end of the last interglacial and ended about 35,000* B.C. *Today it is considered a complex of cultures and/or culture variants, most of which are linked with Neanderthal man.*

Neanderthal: *Fossil type of man first found in the Neander Valley (*thal *in German) of Germany. Squat and strong, he had* *primitive features, which are very often exaggerated by modern writers. He is thought to be an early form of* Homo sapiens, *and he contributed perhaps partially to the formation of modern man.*

Neolithic: *Ensemble of postglacial cultures characterized by agriculture, pottery, domestication of animals, polished stone tools, and villages. Some of these features may be absent.*

Palynology: *The study of pollens and spores of plants.*

Perigordian: *The first of the Upper Palaeolithic cultures in France, characterized by backed blades.*

Pleistocene: *The first and by far the longest subdivision of the Quaternary; began at least 2 million years ago and ended with the Würm glaciation. The question of the boundary between the Pleistocene and the Pliocene (the last of the Tertiary subdivisions) is subject to discussion.*

Potassium-Argon dating: *A method of dating, similar to radiocarbon dating, based on the radioactivity of potassium but applicable only to potassium-bearing sediments. It has a far longer range than the radiocarbon method but is not always reliable for periods younger than 500,000 years.*

Quaternary: *The last of the great geological eras. We live in its second subdivision, the Holocene.*

Radiocarbon dating: *A method of absolute dating based on the radioactivity of an isotope of carbon (carbon 14) formed in the atmosphere by cosmic ray action and assimilated by plants at the same time as normal carbon (carbon 12). It passes then to herbivores and carnivores and can be found in their flesh and bones. After the death of the animal or plant no more C-14 is added, and since this C-14 decays by radioactivity at a constant rate, datation is possible, assuming a constant proportion of C-14 in the atmosphere.*

Retouch: *The shaping of a tool (from a flake or blade) by removing small secondary flakes either by percussion or by pressure; also, the trace of the small flakes taken off in this fashion.*

Riss: *The third of the classical glacial periods; about as extensive as the Mindel.*

Solifluction: *Soil creeping along the slopes, under conditions of alternate freezing and thawing.*

Solutrean: *A short-lived but vigorous culture of the Upper Palaeolithic in western Europe, mainly in France. It lasted from about 19,000* B.C. *to 16,500* B.C. *and is characterized by very finely worked flint projectile points and knives, usually leaf-shaped.*

Stadial: *A period of time corresponding to an advance of the ice sheets during the Pleistocene.*

Stratigraphy: *The study of the deposition of geological layers; also, their order of deposition.*

Upper Palaeolithic: *Ensemble of prehistoric cultures belonging to the second half of the Würm glaciation and characterized by the presence of modern-type men; began around 35,000* B.C.

Würm: *The last of the classical glacial periods, less extensive than the Mindel or Riss; began about 90,000 to 75,000 years ago and ended about 9,500* B.C.

Bibliography

1. Binford, L. R. and Binford, S. R.: A preliminary analysis of functional variability in the Mousterian of Levallois facies. *American Anthropologist*, 1966, 68, 2, part 2, pp. 238–295.
2. Binford, S. R. and Binford, L. R.: Stone tools and human behavior. *Scientific American*, 1969, 220, 4, pp. 70–84.
3. Bonifay, E.: Les sédiments détritiques grossiers dans le remplissage des grottes. Méthode d'étude morphologique et statistique. *L'Anthropologie*, 1956, pp. 447–461.
4. Bordes, F.: Les gisements du Pech de l'Azé (Dordogne). I, le Moustérien de tradition acheuléenne. *L'Anthropologie*, 1954, pp. 401–432, and 1955, pp. 1–38.
5. Bordes, F.: Mousterian cultures in France. *Science*, Sept. 22, 1961, pp. 803–810.
6. Bordes, F.: Typologie du Paléolithique ancien et moyen. Delmas, Bordeaux, 1961; 2d ed., 1967.
7. Bordes, F.: Acheulean Cultures in Southwest France. Studies in Prehistory, Robert Bruce Foot Memorial Volume, Calcutta, 1966, pp. 49–57.
8. Bordes, F.: *The Old Stone Age*. McGraw-Hill, New York, 1968.
9. Bordes, F.: Os percé moustérien et os gravé acheuléen du Pech de l'Azé II. *Quaternaria*, XI, Rome, 1969, pp. 1–6.
10. Bordes, F., *et al.*: Livret-Guide de l'excursion A-5, Landes-Périgord. Union internationale pour l'étude du Quaternaire, VIIIème Congrès INQUA, Paris, 1969.
11. Bordes, F. and Bourgon, M.: Le gisement du Pech de l'Azé-Nord. Prise de date et observations préliminaires. *Bulletin de la Société préhistorique française*, 1950, pp. 381–383.
12. Bordes, F. and Bourgon, M.: Le gisement du Pech de l'Azé-Nord. Campagne 1950–51. Les couches inférieures à *Rhinoceros Mercki*. *Bulletin de la Société préhistorique française*, 1951, pp. 520–538.

13. Bordes, F., Laville, H. and Paquereau, M.-M.: Observations sur le Pleistocene supérieur du gisement de Combe-Grenal. *Actes de la Société linnéenne de Bordeaux*, 1966, pp. 1–19.
14. Bordes, F. and Prat, F.: Observations sur les faunes du Riss et du Würm I. *L'Anthropologie*, 1965, pp. 31–45.
15. Bordes, F. and Sonneville-Bordes, D. de: The significance of variability in Paleolithic assemblages. *World Archaeology*, June 1970, pp. 61–73.
16. Boriskovski, P. I.: Paleoliticheskaia Stoianka Valukinskovo. Ocherki po Paleolitu basseina Dona. *Materialii i issledovania po Arkeologii SSSR*, No. 121 (Figs. 107–108).
17. Capitan, L. and Peyrony, D.: Deux squelettes humains au milieu de foyers de l'époque moustérienne. *Revue de l'Ecole d'Anthropologie de Paris*, 1909, pp. 403–409.
18. Cheynier, A.: *Jouannet, grand-père de la Préhistoire*. Imprimerie Chastrusse, Praudel et cie, Brive, 1936.
19. Feaux, M.: Grottes du Pey de l'Ase et de Combe-Grenal. Lettres de l'abbé Audierne à de Mourcin. *Bulletin de la Société historique et archéologique du Périgord*, March–April 1908, pp. 121–131.
20. Ferembach, D., *et al.*: L'Enfant du Pech de l'Azé. Archives de l'Institut de Paléontologie Humaine, Mémoire 33, Paris, 1970.
21. Howell, F. C.: *Early Man*. Time-Life Books, New York, 1965.
22. Laïss, R.: Ueber Höhelensedimente. *Quartär*, 1941, pp. 56–108.
23. Lartet, E. and Christy, H.: Cavernes du Périgord. Objets gravés et sculptés des temps préhistoriques dans l'Europe occidentale. *Revue archéologique*, Paris, 1864, pp. 234–267.
24. Laville, H.: Recherches sédimentologiques sur la paléoclimatologie du Wurmien récent en Périgord. *L'Anthropologie*, 1964, pp. 1–48 and 219–252.
25. Lumley, H. de, *et al.*: Une cabane acheuléenne dans la grotte du Lazaret (Nice). *Mémoires de la Société préhistorique française*, 1969.
26. Paquereau, M.-M.: Etude palynologique du Würm I du Pech de l'Azé (Dordogne). *Quaternaria*, Rome, XI, pp. 227–236.
27. Peyrony, D.: Le Moustier. *Revue Anthropologique*, Paris, 1930.
28. Schmid, E.: Höhlenforschung und Sedimentanalyse. *Schriften des Institutes für Ur-und Frügeschichte des Schweiss*, Vol. 13, 1958.
29. Vaufrey, R.: Le Moustérien de tradition acheuléenne du

Pech de l'Azé (Dordogne). *L'Anthropologie*, 1933, pp. 425–427.

30. Vértes, L.: Untersuchungen an Höhlensedimenten. Methode und Ergebnisse. Magyar Nemzeti Muzeum, Budapest, 1959.
31. Warren, H.: The Clacton flint industry, a new interpretation. *Proceedings of the Geologists' Association.* London, 1951, pp. 107–135.

INDEX

72 73 74 75 76 9 8 7 6 5 4 3 2 1